STRATEGIC RIFLEMAN

OTHER TACTICS/INTELL. SUPPLEMENTS FROM POSTERITY PRESS

GUNG HO: THE CORPS' MOST PROGRESSIVE TRADITION
GLOBAL WARRIOR: AVERTING WWIII
EXPEDITIONARY EAGLES: OUTMANEUVERING THE TALIBAN
HOMELAND SIEGE: TACTICS FOR POLICE AND MILITARY
TEQUILA JUNCTION: 4TH-GENERATION COUNTERINSURGENCY
DRAGON DAYS: TIME FOR "UNCONVENTIONAL" TACTICS
TERRORIST TRAIL: BACKTRACKING THE FOREIGN FIGHTER
MILITANT TRICKS: BATTLEFIELD RUSES OF THE ISLAMIC MILITANT
TACTICS OF THE CRESCENT MOON: MILITANT MUSLIM COMBAT METHODS
THE TIGER'S WAY: A U.S. PRIVATE'S BEST CHANCE FOR SURVIVAL
PHANTOM SOLDIER: THE ENEMY'S ANSWER TO U.S. FIREPOWER
ONE MORE BRIDGE TO CROSS: LOWERING THE COST OF WAR
THE LAST HUNDRED YARDS: THE NCO'S CONTRIBUTION TO WARFARE

STRATEGIC RIFLEMAN

KEY TO MORE MORAL WARFARE

ILLUSTRATED

H. JOHN POOLE
FOREWORD BY
GEN. ANTHONY C. ZINNI USMC (RET.)

POSTERITY
PRESS

Published by Posterity Press
P.O. Box 5360, Emerald Isle, NC 28594
(www.posteritypress.org)

Cataloging-in-Publication Data
Poole, H. John, 1943-
Strategic Rifleman
   Includes bibliography and index.
   1. Infantry drill and tactics.
   2. Military art and science.
   3. Military history.
I. Title.   ISBN: 978-0-9818659-5-9   2014   355'.42
Library of Congress Control Number: 2013920662

Coverart composition © 2014 by Posterity Press
Edited by Dr. Mary Beth Poole
Proofread by William E. Harris

First printing, United States of America, June 2014

For future U.S. military inductees, that they might have some chance of reaching their full warrior potential.

Taps

DAY IS DONE.
GONE THE SUN.
FROM THE LAKES
FROM THE HILLS.
FROM THE SKY.
ALL IS WELL.
SAFELY REST.
GOD IS NIGH.

FADING LIGHT.
DIMS THE SIGHT.
AND A STAR.
GEMS THE SKY.
GLEAMING BRIGHT.
FROM AFAR.
DRAWING NIGH.
FALLS THE NIGHT.

THANKS AND PRAISE.   FOR OUR DAYS.
NEATH THE SUN.
NEATH THE STARS.
NEATH THE SKY.
AS WE GO.
THIS WE KNOW.
GOD IS NIGH.

A "strategic rifleman" would never die in vain.

Contents

CONTENTS ————————————————————————

Part Five: Properly Preparing the New Squad Member

Chapter 14: ALL-HANDS ACCOUNTABILITY FOR MORE MORAL UNITS 185

Each rifleman trusted with strategic mission

Chapter 15: REESTABLISHING INDIVIDUAL INITIATIVE 193

Strict rules for when Privates may question procedure
Guerrilla warfare training an essential ingredient
Squad management by exception after guidelines

Chapter 16: PERSONAL-DECISION-MAKING PRACTICE 207

Decision-oriented Gurkha Trail
Advanced fire-and-movement course
Crawling for distance and speed
Low-trajectory grenade throwing
The urban-survival competition
Battledrills versus situation stations

Chapter 17: TROOPS MUST HELP TO DESIGN OWN MOVES 233

Tactical "Gung Ho Session"
Bottom-up company quest for better squad technique

Afterword: *NO MINOR OVERSIGHT* 243
Appendix: *KOREAN WAR SIGHTING* 251

Notes

Source Notes 257
Endnotes 261

Glossary 275
Bibliography 279
About the Author 287
Name Index 289

x

Illustrations

Chapter 14: *All-Hands Accountability for More Moral Units*

Chapter 15: *Reestablishing Individual Initiative*

Chapter 16: *Personal-Decision-Making Practice*

Chapter 17: *Troops Must Help to Design Own Moves*

Afterword: *No Minor Oversight*

Appendix: *Korean War Sighting*

Tables

Foreword

Two and a half decades ago, at the end of the Cold War, U.S. military leaders and strategists began talking about a military "transformation". We would redesign a military for the 21st Century that would leverage technology and innovation. With an all volunteer military created in the aftermath of Vietnam and the elimination of the threat of a war on Europe's plains, it seemed an opportune time to rethink our military.

Unfortunately, the path toward transforming the military began with the assumption that ground forces were becoming obsolete and wars could be won through stand-off high-tech systems that allowed detection and engagement at safe distances from harm's way. Like in a video game, we could fight our wars in a sanitary and detached manner. The 21st Century battlefields we encountered, however, frustrated our attempts to make this technology dream a reality.

Instead of cutting ground forces as a bill payer for all this advanced technology as intended, conflicts such as Somalia, Iraq, and Afghanistan forced reluctant increases in ground forces (Army and Marine Corps) to meet the demands of boots on the ground. We still needed troops physically there to control terrain and people, and to engage an enemy that was not completely susceptible to defeat through our technology attacks.

Despite the lesson that ground forces are still relevant and required on today's battlefield, we stubbornly remain committed to drastically cutting them in favor of the technology solutions. Even political leaders who advocate intervention, are quick to add the politically correct caveat, "but no boots on the ground".

What John Poole has creatively demonstrated in all his books, and particularly in this publication, is that the answer is not to eliminate ground forces but to make them more capable. We have seen how effective our Special Operations Forces have become in the current fights we are engaged in. John rightly contends that investment in skill development, leader education, and innovative training can produce far more effective ground forces that can operate independently in small unit formations.

There is a gap between the few highly skilled SOF units and the high-tech capabilities we are developing. That gap has grown larger as we refuse to invest in our ground forces and make them more capable to meet the demands of today's battlefield. John Poole offers a unique and creative solution to fill that gap in *Strategic Rifleman.*

GEN. ANTHONY C. ZINNI USMC (RET.)
FORMER HEAD OF CENTCOM

Preface

The Still Pertinent Historical Setting

Don't be lulled into thinking that United States (U.S.) ground forces have already added—to their "corporate knowledge"—all that history and experience had to offer. Big bureaucracies would rather perpetuate existing procedure. War is a continuum, and many aspects—short-range combat in particular—change very little with most technological advances. That's why present-day U.S. gladiators must continually revisit to the past.

Early in 1999, U.S. Marine Commandant Gen. Charles C. Krulak started pushing for a "Strategic Corporal." Not all E-4's of every occupational specialty were to have this increase in responsibility, only the infantry squad leaders.[1] Each of their squads was to be somehow readied to single-handedly influence the overall war effort. Of course, such a thing would be more likely if every squad member were first aware of strategic goals, and then well enough trained to personally contribute. In other words, the key to Gen. Krulak's vision for how better to fight in the 21st Century was a "Strategic Rifleman." To confirm this, one must look back into history.

After running out of ammunition and still managing (through a bayonet charge) to defend Gettysburg's Little Round Top, Col. Joshua Chamberlain tried to convince all men of their "divine spark." True military professionals hate war and yearn for ways to more morally conduct one. The Pentagon's favorite (and economy-enhancing) avenue has been through ever-more-powerful standoff bombardment. But, bombardment (however precise) still results in collateral damage. While theoretically removing all need for ground involvement, "death from above" is unable to win most wars. Though partially interdicted, the battlefield is still under opposition control. If such a conflict is eventually to be abandoned, then all that goes into its prosecution becomes unethical. To be on a firm moral footing and assured of winning, one must find some way to occupy the contested ground. Most high-level American bureaucrats believe a large military contingent with far-reaching support apparatus necessary. Since the Vietnam War, they have been afraid to decentralize enough control to blanket a vast area with token forces. Only after too little small-unit preparation should such a strategy become unacceptably dangerous. Yet, even America's special-operations community has been struggling with its prerequisites.

Combat Ethics an Important Part of Modern Warfighting

Wartime morality is most logically enhanced through less firepower (of whatever accuracy) and more surprise-generating maneuver. The latter takes squad autonomy to a degree that most "top-down" organizations find uncomfortable. What has resulted throughout the U.S. officer corps is a gross underestimation of what a single American "gyrene" or "dog face" can accomplish on his own in combat.[2] Their politically rooted desire to avoid all casualties has produced destruction-dispensing drones, but no real change to the rifleman's "premachinegun" ways of traversing a battlefield. Those outdated movements and their squad combinations have already resulted in several small-war losses and too many big-war casualties.

Masking this system-wide "fall" from tactical excellence has been no shortage of misplaced "pride." The American officer community is not being paid to perpetuate existing organizational structure, but to win wars. Through it all, each U.S. nonrate's spiritual health has been no better protected than his warfighting potential.

Of all the battlefield issues, ethics is the most difficult to discuss. That's because many consider war so inherently evil that strictly obeying any moral edict reduces the chances of victory. Instead, one must repay enemy excesses in kind and spare frontline soldiers from any introspective diversion. While their point is well taken, it is not the all-encompassing answer to every wartime dilemma. If that soldier has a "soul," then shouldn't his headquarters try as hard to protect that soul as his physical body? Whatever it decides to do in that regard could simply be added to his rehearsal regimen, like no longer allowing him to fire blanks into a hand-raised "aggressor." After acquiring all appropriate habits, that nonrate would then be less likely to hesitate during close-quarters combat. Every action—moral or otherwise—would simply be part of his muscle memory.

Much has been said about "just wars." The Roman Catholic Church has even listed their parameters. But, within the American military establishment, the whole issue of "combat morality" has been generally skirted. Two explanations are possible: either (1) war is so inherently immoral that all participants must be allowed to do whatever it takes to survive; or (2) combat "technicians" automatically shy away from anything philosophical. One cannot blame these frontline veterans for resisting behavioral limits. Mortal combat happens quickly. In the split second it takes to consider the ethical ramifications of one's next act, the enemy can get the upper hand. To a person fighting for his very life, all options seem legitimate. Unfortunately, some of those options result in unseen wounds that may later diminish parent-unit effectiveness. Throughout World War II (WWII), such emotional injuries were all lumped into the "Combat Fatigue" category. Now, they fall under "Post Traumatic Stress Disorder (PTSD)." (See Figure P.1.) However effective the individual was before their occurrence,

Figure P.1: Combat Can Inflict Invisible Injuries
(Source: "After the Battle," by Michael R. Crook, U.S. Army Center of Military History, posted June 2007, from this url: http://www.history.army.mil/art/A&I/0607-2.jpg)

he (or she) now suffers debilitating consequences. As such, one can make a good case for more headquarters' attention to each combatant's mental and spiritual health. In effect, their "souls" have become just as important to winning the overall conflict as their physical well-being; not to mention the fact that intact souls might less often commit suicide.

Yet, one is irresistibly drawn back to the first possibility for why "combat morality" has been underemphasized in America— unavoidable responses to desperate circumstances. While "total war" can be fairly nasty business, modern wars are of a much lower intensity. All have economic, political, and religious overtones. Within such a conflict, the idea of completely destroying a nation's infrastructure to more easily free its people becomes nonsensical. (See Figure P.2.) As in law enforcement, too much collateral damage will simply alienate the population. Tiny loosely controlled maneuver elements seem to get the most done under certain conditions—just like police "Bobbies" on a beat. Among those conditions are enough team proficiency to operate semi-autonomously and team dispersion to permit mutual assistance.

Figure P.2: Too Much Collateral Damage Now Loses Wars
(Source: "St. Lo Patrol," by Olin Dows, U.S. Army Center of Military History, from this url: http://www.history.army.mil/art/dows/4_229_46.JPG)

One Need Not Be a Communist to Use Red Army Methods

    Within the Asian Communist way of employing infantrymen lies a nontraditional solution to the Pentagon's now chronic problem. Though lately associated with authoritarian government, this infantry option best approximates the law enforcement model. Instead of always marching to the sound of the guns and then expecting subordinates to win every engagement, the Asian commander looks for situations in which his people are sure to excel. This more realistic (though less charismatic) way of leading creates a different mind-set for all frontline fighters. (See Figure P.3.) Less often patronized, each Asian rifleman must instead face up to his own inadequacies. To help him compensate, he is shown more ways to

Figure P.3: Not the Mindless Automaton of "Cold War" Hype
(Source: Courtesy of Cassell PLC, from "World Army Uniforms since 1939," © 1975, 1980, 1981, 1983 by Blandford Press Ltd., Part II, Plate 80)

keep from getting hurt. Killing is no longer his mission. Whenever possible, he avoids physical contact to more easily get at the foe's strategic materiel. Not only has he been entrusted with part of the overall mission, but he is less often asked to override his Creator's design. He thus becomes more self-contained than his Western counterpart.

Asian soldiers have no unique claim to such a capability. The extent to which U.S. service personnel have endured organizational inertia over the years makes them every bit as patient. Within the Asian rifleman's training simply lies the model for solving the Pentagon's most pressing problem—how "decisively" to win another war. Though quite painful to admit, America hasn't done so since WWII.

LT.COL. H. JOHN POOLE USMC (RET.)
FORMER GUNNERY SERGEANT FMCR

Acknowledgments

There are no new ideas within the U.S. military, only former discoveries that were never adopted. One need not go to Asia to find the keys to better infantry performance, only reassess America's own war record. Unfortunately, a Western-style bureaucracy tends to resist any change, however useful. That's why *ninpo* (advanced Japanese *ninjutsu)* justification was deemed necessary for a few "outside-the-box" recommendations.

Part One

The Supernatural Legacy of Armed Conflict

THE HUMBLEST CITIZEN . . . , WHEN CLAD IN THE ARMOR OF A
RIGHTEOUS CAUSE, IS STRONGER THAN ALL THE HOSTS OF ERROR.
— WILLIAM JENNINGS BRYAN

(Source: Attributed to William Jennings Bryan.)

Evil Has Long Influenced War

- Do only honorable deeds occur in battle?
- Which well-intended decisions have led to wrongdoing?

How a "good" wartime angel might now look.

Evil in Combat

In what is largely violent by definition, the concept of evil only has much meaning in a religious context. That which too badly violates God's laws might be considered "evil" if the perpetrator were fully aware of the restriction. To be on a firm moral footing, the Christian soldier need not abstain from all killing. He must only limit that effort to enemy combatants and then show each as much mercy as possible.

As America entered World War I (WWI), Sgt. Alvin York overcame his inner conflict over Jesus's advice to "love your enemies" with another Biblical passage: "Render . . . unto Caesar the things that are Caesar's, and to God the things that are God's." During the Meuse-Argonne Offensive of 1918, he and two subordinates then killed 20 Germans and captured 100 more to save their own patrol from annihilation. (See Figure 1.1.) When asked why he did so much lethal shooting, York said: "To save lives."[1] That's quite different from responding in kind to enemy evil.

Low-ranking American infantrymen are thought to be just following orders, and so normally excused from any stigma of religious wrongdoing. Yet, the need for those orders to be "lawful" still creates a certain amount of inner tension. Lawful is a word generally taken to mean "in compliance with the Geneva Conventions." Yet, Washington has yet to consider any insurgent worthy of full protection under those Conventions (having never ratified the 1977

Figure 1.1: Sgt. York Was Morally Conflicted in WWI
(Source: http://search.usa.gov public-domain image from this url: www.history.army.mil/images/artphoto/pripos/wwi/infantryman.jpg)

4

Figure 1.2: Americans Had Been Forced to Kill One Another
(Source: "Plenty of Fighting Today," by Keith Rocco, poss. ©, from this url: www.nationalguard.mil/resources/photo_gallery/heritage/hires/Plenty_of_Fighing_Today.jpg)

Protocols).[2] That may help to explain why—after closely obeying all command directives—so many veterans of guerrilla wars come home with inner conflicts.

The Enlisted Consensus on War

While some U.S. Government Issues (GIs) believe in the "Devil" and some do not, all agree that war can be extremely distasteful at short range. Among its greatest abominations has been the killing of politically naive soldiers over some error in diplomatic or strategic judgment. Just before the Battle of Stone's River during the early stages of the American Civil War, both armies sang "Home Sweet Home" in unison.[3] Then, neighbors and cousins had to slaughter one another. (See Figure 1.2.)

Most wars are not won through killing, but by limiting the adversary's source of resupply. That may be why "evil" starts to creep into the picture as soon as more enemy fatalities become equated with quicker overall victory. Pope John Paul II alluded to this in his *dictum* on national defense. To be most ethical in war, one must try to limit opposition casualties.

> In order to be legitimate, the "defense" must be carried out in a way that causes the least damage and, if possible, saves the life of the aggressor.[4]
> — Pope John Paul II

First Casualty Must Not Be Strategic Reason

Personnel are more easily replaced than materiel. That's why—in a widespread conflict—the relative number of casualties is seldom the deciding factor. It is in a moment of anger, caution, or hurt that too much significance gets attributed to killing. On offense, the argument goes something like this: "Any defender who escapes this attack may be encountered on the next ridgeline." A similar oath—"Kill them all"—can be heard after too many losses on defense. Among its best known proclaimers is "Stonewall" Jackson.

> Dr. McGuire . . . put a question (to Jackson) as to the best means of coping with the overwhelming numbers of the enemy. "Kill them, sir! Kill every man," was the reply of the stern soldier.[5]
> — quote from General Thomas Jonathan Jackson after losing a comrade to enemy assault

If the mighty Stonewall can make such an egregious error in leadership, then so too can any junior contemporary—whether commissioned or not. Some American inductees fail to learn the difference between right and wrong at home or school. They should never be subjected to such a statement in the military. For them, any authority to kill could lead to an excess. Many things have contributed to the wartime adage: "The road to hell is paved with good intentions."

Way of Fighting Really Does Matter

To further complicate matters, America's traditional combat style—Attrition Warfare—has always targeted enemy troop concentrations. Their composite parts have been assigned so little intrinsic value as to be regularly dehumanized by propaganda. Yet those parts are still children of God, albeit misguided. By opting for firepower over maneuver, the Pentagon may have unintentionally encouraged too little regard for human life.

Such a conclusion won't please many U.S. war veterans. They have only done what they were ordered or compelled (by survival instinct) to do. Many are unaware of the 1980 United Nations (U.N.) Convention on Certain Conventional Weapons (CCW) to "prohibit the use of weapons that are indiscriminate or of a nature to cause superfluous injury or unnecessary suffering."[6] Others seem oblivious to the much older Geneva Convention requirement that a hand-raised defender be allowed to surrender.

In retrospect, many former U.S. infantrymen would have much rather obeyed God's laws than responding in kind to any enemy excess. Still, that's the unfortunate nature of war. Quite possibly, killing so conflicts with man's inner (God-given) nature, that even the fully justifiable kind causes it damage. That would certainly explain the venerable claim that "there are no true winners among those who must [actually] fight a war." Yet, a way to limit this psychological and spiritual toll must exist. Varying the degrees of training, supervision, maneuver, and firepower are all possibilities. Only missing is a specific proposal.

The Very Real Presence of Evil in Life

Whether or not the average U.S. rifleman worries about "demonic" pressures in combat, the world's major religions still do for every aspect of human endeavor. Since their inception, all have considered mankind to be continually under attack by some personification of evil.[7] Each American soldier or Marine should therefore be interested in avoiding its influence. At the squad level, that would most logically take enough proficiency and autonomy to occasionally show mercy in combat. However, that much leeway is seldom allowed from a headquarters that considers itself the only road to propriety. Though tight overall control may at times stifle local atrocity, it can also lead to widespread error—like the poisoning of an entire population (as well as one's own troops) through the overuse of a defoliant.

All U.S. headquarters do exercise some battlefield discretion through their "Rules of Engagement," but most have yet to acknowledge any correlation between morality and maneuver at the squad level. Maneuver (at any echelon) brings more surprise and enemy capitulation, whereas its alternative—firepower—can only lead to additional death. Tactical opportunism requires initiative, and most commands worry that too much initiative at the lower echelons would cause vital orders to be questioned. Thus occurs the ongoing debate over how best to control things on what should be a "field of honor." That more self-sufficient riflemen could accomplish the same mission with less collateral damage will take many chapters to prove. But first, one usefully recalls how evil has affected other armies in history.

Things That Went Bump in the Night

In times of trouble, the ancient Chinese believed they could summon supernatural warriors from the Netherworld.[8] Good angels have periodically appeared on Western battlefields, so why not a few of the fallen variety? After all, "resisting the Devil's tactics . . . and spiritual army of evil" does constitute the Christian definition of Spiritual Warfare.[9]

According to the Bible, up to one third of "millions" of angels rebelled with Lucifer and then fell from heaven.[10] One of the most disturbing legends to come out of the American Civil War was that demons poured out of the ground in 1864. Possibly based on the subhumans who went around killing the wounded after the Battle of the Wilderness, this legend still suggests more of an opportunity for evil as any war intensifies. The Bible also implies that each person has an "unfriendly" angel (as well as their guardian) nearby at any given time. Such a presence would require more caution on everyone's part.

I was given a thorn in the flesh, an Angel of Satan to beat me and stop me from getting too proud! About this thing,

I have pleaded with the Lord three times for it to leave me, but He has said: "My grace is enough for you: my power is at its best in weakness."[11]
— 2 Corinthians, 12:8

God's still obedient "messengers" are more fun to watch. They protected the Israelites twice,[12] and once even helped Pope Leo I to save Rome.

When notorious warrior Attila the Hun and his massive army tried to invade Rome during the year 452, Pope Leo I met with Attila to plead with him to stop threatening Rome. Many people were surprised that, in response, Attila immediately withdrew his army from Rome. Attila said he left the city because he saw two imposing angels wielding flaming swords standing beside Pope Leo I while he was speaking.[13]

As Allied forces were about to be overrun at the Battle of Mons in the early stages of WWI, another "good" angel appeared. Brigadier General Charles Charteris later wrote the following: "[T]he Angel of the Lord . . . , clad all in white with flaming sword, faced the Germans and forbade their further progress."[14] This may be one and the same with the Korean War entity in the Appendix.

Wartime Atrocities

History's more lengthy conflicts all have involved massacres of fighters and civilians alike, but the modern concept of "atrocity" is most often associated with America's WWII adversaries. (See Figure 1.3.)

German *Schutzstaffel* (SS) Battalions are known to have conducted hundreds of mass executions, but Japanese infantrymen are more often accused of illicit behavior. Largely at fault is the *samurai* code of *"bushido"* that they universally adopted after a lackluster performance in the Russo-Japanese War. From this code evolved a complete disdain for any enemy who would not fight to the death. Their "valuing [of] honor above life"[15] subsequently led to great dishonor (and indeed evil) through the maltreatment

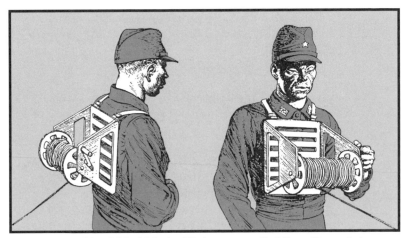

Figure 1.3: "Bushido" Made All Japanese Soldiers Seem Inhuman
(Source: "Handbook on Japanese Military Forces," TME 30-480 [1944], U.S. War Dept., p. 321)

of prisoners. This same Japanese code made it much more difficult for U.S. personnel to maintain a high moral standard during their hard-fought campaigns in the South Pacific. (See Figure 1.4.)

Headquarters' Intercession

Most Westerners must believe that junior enlisted personnel require close supervision to avoid barbarous acts in combat. Such a viewpoint fits in comfortably with their "top-down" management tradition. Yet, one U.S. infantry commander claims his lowest-ranking troops provided him with the "best moral compass" in Vietnam.[16] Thus, the average ethics of unit members—when combined with a little peer pressure—may also result in totally acceptable behavior.

This "troop-friendly" argument is most often rebutted with the enemy excesses of WWII. Wherever there was a tiny semi-autonomous Japanese contingent, bad things seemed to happen to local inhabitants and Allied hostages alike. The reason was automatically assumed to be too little control. The effect of *bushido* on group behavior was never considered. Might over-exuberance with any parent-unit "toughness" campaign lead to the same type of problem?

The answer to that question may lie in what happened at My Lai in Vietnam. Though variously described as "group evil," it was little different from a U.S. Indian War "massacre." The latter term has since been equated with poorly disciplined or morally deficient troops. More likely, "body counts" had been so highly touted as to cause the assault unit commander's "frag order" to be misinterpreted. Thus, one wonders how the My Lai platoon might have done with more peer pressure, and less parent-unit influence.

Figure 1.4: Pacific Theater GIs Had Trouble Taking POWs
(Source: "Red Arrow at War," by Michael Gnatek, poss. ©, from this url: www.nationalguard.mil/resources/photo_gallery/heritage/images/redarrow.jpg)

Figure 1.5: 2nd Raider Battalion Arm Patch
(Source: Wikipedia Encyclopedia, s.v. "Carlson's Raiders," with image designator "Marine_Raiders_insignia.svg")

GIs Caught between a Rock and a Hard Place

Evil is an affliction so seldom total as to be easily rationalized away. That's why Christians so readily admit to being sinners, and the Devil is called the "great deceiver." This affliction becomes more likely where some religious warning has been ignored—e.g., "Refrain from vengeance."

For people to avoid any occasion of evil, they would have fully to assess each act, its associated circumstances, and its later consequences. Few American military leaders allow their bottom-echelon fighters that much introspection. It might lead to too many command directives being overlooked. To avoid moral error, the rank and file must wholly rely on any Rules of Engagement and the assumed legality of their orders.

Only on Occasion Have U.S. Troops Enjoyed More Leeway

For a while during the early stages of WWII, Lt. Col. Evans Carlson got away with an "Ethical Indoctrination" of his Marine Raiders. Unfortunately, its content may have too closely followed Mao's model. Though Mao greatly respected his individual fighters as instruments of war, he readily expended them. With little personal regard for human life, he only spared enemy hostages as a psychological ploy.

Carlson had used this Indoctrination to inform his fighters of a just cause, not the moral pitfalls of combat. As suggested by the 2nd Battalion patch, his ultimate goal was still attrition. (See Figure 1.5.) Then, he made the mistake of searching out those who didn't mind pursuing that goal with a knife. While killing remained an unfortunate part of Attrition Warfare and close-quarters knife work useful to commandos, Carlson may have too widely opened the door to another issue. Those too comfortable with a messy killing might be less adverse to excess killing. Soon, his "leadership by consensus" experiment began to unravel. When one of his Raiders was captured and tortured to death at the beginning of the Long Patrol on Guadalcanal, the good Colonel became conflicted as to what to do with subsequent Japanese detainees. Not much of a Prisoner of War (POW) pipeline existed at that time; and badly outnumbered guerrillas must travel light. As confirmed by the official chronicle, Carlson then turned his first hostages over to the dead Marine's friends to "take care of."[17] The predictable happened, and with it a fatal crack in any moral foundation. As the war intensified and more enemy soldiers refused to surrender, this predisposition toward the "elimination of all resistance" must have spread throughout the Fleet Marine Forces (as suggested by the historical drama, "Pacific"[18]). To what extent, this overemphasis on killing may have undermined the war effort, no one really knows. Only certain is that a lot of young men—with every right to be immensely proud of their accomplishments—came home with bad dreams. In all probability, the same thing had happened to U.S. Army survivors of the hard-fought European campaigns.

The Inescapable Conclusion

If warfare really does present more of an opportunity for evil, then every participant (especially, those in closest contact with the enemy) should be required to reach a certain level of self-discipline.

> All that is necessary for the triumph of evil is that good men do nothing.[19]
> — Edmund Burke (British Statesman
>   who supported the American Revolution)

Mystical Tricks of a More Worldly Origin

● Have U.S. sentries ever been subjected to enemy hypnosis?

● What form did it take?

A few woods dwellers do sound like ghosts.

More Recent "Extra-Normal" Phenomena

Though the ancient Chinese may have summoned Netherworld warriors, slightly more modern Persians conjured up dust-like impediments to visibility. Such a cloud made it possible to sneak up on a desert quarry in broad daylight. In the late 13th Century, Marco Polo warned of a daytime "darkness" being somehow generated by the caravan-raiding *Karaunas* of Upper Persia.[1] Upon retracing Marco Polo's steps, Col. Henry Yule and Maj. Sykes then encountered a "dry mist" in Upper India.[2]

Though these historical sightings have been mostly ignored,

present-day thinkers are now faced with a videotaped record of white flakes floating around a darkened room. Investigators of the paranormal attribute such flakes to ectoplasm.[3] So, today's warriors must keep an open mind to all they don't fully understand. Truth can be far stranger than fiction.

Just prior to the famous "vanishing of the Sandringham Company" at Gallipoli in 1915, a strange haze was hanging over the British unit's attack objective. In the high brush and deep-ditches that covered the arid flatlands behind Sulva Bay, that unit's rank and file were last seen assaulting on line into a cloud of "etherial yellow mist."[4] Then, all traces of them vanished. Subsequent studies have attributed their disappearance to everything from spaceship abduction to witchcraft. More likely, Turkish irregulars were hiding in row of spider holes to shoot every British infantryman simultaneously in the back from a self-imposed smoke screen. Then, the poorly outfitted Turks grabbed all the British "kits" and hid their owners' bodies as a form of intimidation.

The Clouding of Men's Minds

While airborne particulate matter may be at the heart of both Persian and Turkish mysteries, there is another Eastern legend that entails mentally blocking an adversary's sensibilities. Early radio episodes of *The Shadow* claim "the clouding of men's minds" to be of Arab origin.[5] Others and the movie remake attribute it to inhabitants of the Tibetan region. As Mongolians occupied much of South and Southwest Asia for centuries, this trick's ultimate source may be well east of Tibet. (See Figure 2.1.)

Clouding another's mind is quite different from the personal "invisibility" described in the *Bhagavad Gita, Visuddhimagga* ("Path of Purification"), and *Tibetan Book of Living and Dying*. That invisibility mostly resulted from emptying one's own mind of any hostile intent.

Within the contemporary world, the subtle altering of another human being's perceptions would more likely come from some form of "hypnosis" than supernatural empowerment. For this reason, the wartime history of hypnosis must next be examined. Should one exist, the Pentagon may need to better prepare its perimeter guards.

Figure 2.1: Sun Tzu Ruses Predate This Soldier by 2150 Years
(Source: Courtesy of Stefan H. Verstappen, from "The Thirty-Six Strategies of Ancient China," © 1999 by Stefan H. Verstappen)

A Military Application of Hypnosis?

Must the recipient of all hypnosis fall into a sleep-like trance, or might a lesser form leave him fully awake but less vigilant? The latter would better facilitate the Oriental soldier's predisposition toward nighttime short-range infiltration (sneaking through enemy lines). And what of the daytime sentry who is openly approached by a non-uniformed Asian? Might that sentry's senses be somehow dulled as well? Pentagon planners should be just as interested in these questions as recent inductees.

Over the years, the Oriental art of war has paid much more attention to frontline perceptions than its Western counterpart. That's why Asian armies have more ways to trick opposition soldiers. Some of those ways (often with Chinese roots) have resulted in long-standing and detailed regimens. For example, *ninjutsu* is the art of stealth through some form of physical accomplishment; and its cousin, *ninpo,* is the "way of invisibility" through enlightenment (the utilization of universal laws). Thus, *ninpo* is considered to be a higher and more spiritual form of *ninjutsu* for all circumstances and levels of human activity.[6] Through spiritual study, the *ninpo* practitioner tries to capitalize on those elements of everyday life that are normally attributed to "luck," "coincidence," or "fate."[7] For the Western proponent, too little understanding of *ninpo* could lead to an ethical miscalculation.

In essence, *ninpo* combines *ninjutsu* with *yojutsu* (the "mystical") and *genjutsu* (illusion).[8] While the Far East might be expected as their point of origin, both are actually metaphysical disciplines from the Indian subcontinent.[9] Based on Himalayan Tantric (spiritual) teachings,[10] *ninpo mykko* helps its practitioner to better focus through meditation.[11]

While the previously mentioned Hindu texts do contain a formula of sorts for how not to exist,[12] their pertinent passages mostly cover how to so empty one's own consciousness as to avoid telegraphing any agenda. Within the *ninpo* version of becoming invisible *(on-shinjutsu),*[13] is more focus on shapes and images with which to hide from a foe. In the process, one can do what any encircled special operator would love—create a physical opening (or exit route) where none previously existed. As Table 2.1 definitively verifies, the *ninpo* practitioner is also able to "cloud the mind"—through hypnotic suggestion—of anyone who gets in his way.

Other *Ninpo* Ways to Alter a Foe's Degree of Awareness

Ninpo techniques for "clouding a sentry's mind" fall under the broad category of *saiminjitsu.*[14] Among them is hazing. Monotonous motion has been often associated with hypnosis. Why couldn't any repetitive sight or sound have the same effect on a U.S. watchstander? Only deep hypnosis would require his willingness. Shallow hypnosis is only "an altered state of intense but narrow

Achieving harmony with universal forces [15]
   Protecting one's own body, mind, and spirit [16]
   Developing a benevolent heart (jihi no kokoro) [17]
   Secret spiritual power (mykko, tantra, and ekkyo) [18]
   Viewing reality from the outside (kongokai mandala) [19]
   Viewing reality from the inside (taizokai mandala) [20]
Ascetic power (shugendo) [21]
An intuitive knowledge of fate [22]

Accomplishing one's mission through indirect influence (bo-ryaku) [23]

Psychological strategies (from yojutsu and genjutsu) [24]
   Spiritual refinement (seishin teki kyoyo) [25]
   Achieving a relaxed mental state [26]
      Channeling one's energy through chants & finger weaving (kuji-in) [27]
Mind control (saiminjitsu) [28]
   Harnessing one's subconscious through self-hypnosis [29]
      Maintaining a "can-do" attitude
         Gaining inner strength through meditation [30]
         Transcending fear or pain [31]
      Enhancing one's awareness of surrounding conditions [32]
         Concentrating on sensory impressions [33]
         Paying attention to detail [34]
         Reading the thoughts of others [35]
         Perceiving danger [36]
         Making decisions [37]
         Visualizing the task to be accomplished [38]
   Clouding a sentry's mind [39]
      Hazing the sentry [40]
      Distracting the sentry through hypnotic suggestion [41]
      Paralyzing the sentry with a spell (fudo kanashibari no jutsu) [42]
  Threatening appearance or shouts as a weapon (kiaijutsu) [43]
Scouting methods (sekkojutsu) [44]

Penetrating a fortress
   Falling in behind a sentry or incoming patrol [45]

Hiding (inpo) [46]
   Making oneself invisible (onshinjutsu) [47]

Escaping (tonpo) [48]
      Establishing a false exit (e.g., opening an outside door without leaving) [49]
      Leaving a recognizable trail and then retracing one's steps [50]
      Doubling back after the foe has bypassed one's concealed position [51]
      Abruptly terminating one's trail [52]

Realm devoid of specific recognizable manifestation (ku no seikai) [53]

Table 2.1: A Few of the Ninpo Capabilities

concentration."[54] Thus, a lone American defender might become so alerted to one abnormality within his assigned sector as to be relatively oblivious to all others.

> A popular myth is that hypnosis is a form of sleep. . . . In fact, EEG [electroencephalogram] studies show that the hypnotic state is not a form of sleep at all. It is a form of focused alertness, with increased attention in one area and decreased or absent focus on other events. . . . Hypnosis is perhaps best described as a daydream so intense that one temporarily believes one is in it, and is oblivious to anything else going on in the surroundings.[55]

Russian units have long used feints to the front of opposition sentries so that a few flankers can crawl to within assaulting distance.[56] Why couldn't an Asian infiltrator's accomplice so distract a U.S. foxhole occupant to permit his partner's passage along the hole's other side? Or, by cooing like a morning dove, might a lone *ninja* sapper generate his own transit opportunity? At 4:00 A.M., such a sound might cause an exhausted teenager to "day dream" of home.

Ninjas are, after all, experts at narrowing their own fields of concentration. A well-studied authority on both human psychology and Eastern religions has this to say about their wartime potential.

> Perhaps some form of meditation is involved in which the outer form or appearance is somehow altered so as to render the practitioner invisible to the eye of the beholder [quarry].
>
> When I have stalked animals, I have been able to get within a few feet of deer, and woodchucks and whatever, even when they have been looking right at me. Needless to say I moved very slowly, and tried to blend in with the surroundings. Keeping an inner stillness seemed to help.
>
> Most of us have an inner dialogue going, with one part of ourselves debating with another part . . . about the best way to do something or other. One result . . . is that we are not processing what our outer perceptions are telling us: e.g., wind shifts, noises (that stamp of the hoof when a deer is beginning to get spooked, or a snort that says he/she

doesn't like what the winds or motion are telling him/her).
So by being empty (no inner dialogue), the hunter stays alert
to what his sensory perceptions are telling him and . . . to
premonitions, intuitions, about the prey. He can sometimes
feel its eyes on him and know when to freeze and when to
move.

. . . [S]topping the inner dialogue . . . makes one invisible
because one is more a part of nature at that moment and
not a preoccupied human being. You can't even want to kill
the prey; they will pick that up.[57]
— Dr. David H. Reinke

As for Those Mystical Spells

"Paralyzing the sentry's mind with a spell" has been so widely
practiced in *ninpo* as to deserve its own name—*fudo kanashibari no
jutsu.*[58] (Look at Table 2.1 more carefully.) The word "spell" normally
infers some magical power. Several well-respected *ninpo* masters
claim to be able to hypnotize an adversary in broad daylight.[59] One
Chinese author even shows how to do it. Just to be on the safe side,
all deployed U.S. service personnel should be adequately warned.
Any American sentry who is openly approached (in the Near, Middle,
or Far East) by a lone figure with strangely clasped hands, should
not look too hard at those hands.

> *Kuji Kiri* is the technique of performing hypnotic movements
> with the hands. These magical in-signs created by knitting
> the fingers together may be used to . . . hypnotize an ad-
> versary into inaction or temporary paralysis. Each is a key
> or psychological trigger to a specific center of power in the
> [human] body. . . . These are keyed to the twelve meridiens
> [sic] of acupuncture.[60]

Oh, By the Way

Of particular note on Table 2.1, all practitioners of *ninpo* must
first develop a "benevolent heart" *(jihi no kokoro)*. *Ninpo's* more
earthy cousin does contain a few very lethal sentry removal techniques, but its main focus is still on "self-defense." The *ninjutsu*

Figure 2.2: WWII Forerunners of All U.S. Special Operators
(Source: http://search.usa.gov public-domain image from this url: http://www.ndguard.ngb.army.mil/history/164WWII/douglasburtell/PublishingImages/BurtellBougMtoil.jpg)

specialist tries only to sneak—undetected—into an enemy fortress. That can be most easily accomplished by avoiding all opposition contacts. In contemporary terms, he intentionally bypasses front-line defenders to more easily get strategically important rear-area targets. This makes *ninjutsu* quite different from most Western warfighting disciplines. They want first to eliminate any frontline resistance.

Either Mao's "Ethical Indoctrination" had failed to include "developing a benevolent heart" or Carlson missed it. If he hadn't, his experiment in "leadership by consensus" might have turned out better. (See Figure 2.2.) His Raiders were the forerunners of all U.S. special operators, so their hard-won lessons should not now be ignored. Any group of well-raised young Americans will have difficulty warming up to the idea of killing as their only purpose. Nor should they have to.

Part Two

Modern Wars Spill Over into Spiritual Arena

Don't let evil conquer you, but conquer evil by doing good.
— Romans 12:21

(Source: "New Living Translation" [©2007], as retrieved from following url: http://bible.cc/romans/12-21.htm)

Overwhelming Force No ___ Longer the Answer

● Is there any such thing as a "surgical" airstrike?

● Do other cultures respect this way of fighting?

Has collateral damage ever help to win a war?

(Source: search.usa.gov public-domain image from this url: http://www.nps.gov/history/history/online_books/npswapa/extContent/usmc/pcn-190-003140-01/images/fig10.jpg)

The More Spiritual Nature of Modern War

Though the wars in Iraq and Afghanistan have had Muslim overtones, most Americans still have trouble making any basic connection between war and religion. For them, any suggestion that wars have now spilled over into the spiritual arena in an over-generalization at best—psychological arena perhaps, but not spiritual. The best way to describe the change is more regard for each noncombatant's opinion. In other words, the "hearts and minds" of a beleaguered population now play a bigger role in who wins the conflict.

As religion has always been the champion of such matters, one can logically conclude that modern wars are more easily won by following religious tenets. But, those tenets need not be uniquely Christian. They can be shared by members of other religions or even the unchurched. Among the most prominent is protecting innocent bystanders from harm.

Firepower's Elusive Lesson

The heaviest fighting in U.S. history arguably occurred in May 1945 at the Sugar Loaf Complex that anchored the west end of Okinawa's infamous Shuri Line. When the U.S. Tenth Army could get no farther than its outer approaches, the Marines were called in. After almost two weeks of desperate fighting, they managed—at great cost—to seize a foothold on two of the bastion's three hills: Sugar Loaf and the Horseshoe. Half Moon was still to be seized.

During the (initial) 10-day period up to and including the capture of Sugar Loaf, the 6th Marine Division had lost 2,662 killed or wounded; there were also 1,289 cases of combat fatigue. In the 22nd and 29th Marines, three battalion commanders and eleven company commanders had been killed or wounded.[1]
    — Official Army chronicle of the battle, 2000

In effect, the defenders had so blanketed ideal terrain with below-ground tunnels and above-ground fire as to make it virtually impervious to assault. Two regiments of the 6th Division had to learn this the hard way. Then, it was the 4th Marines' turn. Its 2nd Battalion (successor to the Mao-oriented 4th Raider Battalion) finally did manage—through non-conventional tactics—partially to occupy Half Moon. To do so, it had needed no tank, artillery, or air support. But, the 6th Division commander had already become tired of the carnage. With fire still raining down on Half Moon from outside his lane, he decided to bypass that piece of terrain altogether. Across the other hilltops, his units moved forward. Finally, the city of Naha was invested, and the attack was able to continue.[2]
    By then in the landing, it had become fairly obvious that the

Marines were more interested in maneuver than their brethren in arms. Elements of the Tenth Army were still attacking the center and east end of the Shuri Line, but with their standard "sledge-hammer" (overwhelming-firepower) approach. If a Marine unit hadn't been allowed to enter an Army lane from the side,[3] Shuri Castle might have held out for months.

So, the lesson is clear. The sledgehammer approach doesn't work against a well-situated strongpoint matrix that is fully connected below ground. As 4th Raider Battalion had demonstrated on New Georgia, and 2/4 again at Half Moon,[4] tiny teams of lightly armed personnel can be more effective against such an objective than a fully supported phalanx. That's because only crawling infantrymen can get close enough to mutually supporting bunkers to locate the apertures of, and then suppress fire from, those protecting the forward bastions from either side rear. With no one orchestrating the maneuver, separate fire teams have only to cooperate between lanes and stay roughly on line.

While this evolutionary advance in squad tactics has been largely lost to the present generation, it was briefly repeated some 25 years later.

A More Recent Example

General Westmoreland's "search-and-destroy" sweeps in Vietnam were ineffective for similar reasons. The Viet Cong's (VC's) strongpoints, reinforcement/escape routes, and safe havens were also well buried. Just finding their entrance and exit portals would have taken mantrackers and war dogs in every unit. Yet, inadequately engaging a wily adversary is not the sledgehammer's only drawback.

Firepower Causes Short Rounds and Collateral Damage

Whenever firepower takes priority over maneuver, there will be friendly "short-round" losses and civilians mistakenly targeted. The first statistic is difficult to ascertain for the Battle of Okinawa, but the second isn't. By some estimates, a full third of Okinawa's 450,000 civilian occupants perished in the Allied invasion, with another third being wounded.[5] A few of Okinawa's residents did willingly fight alongside the Japanese, but most had been involuntarily "impressed" as laborers.[6]

Then, all that artillery and air bombardment in Vietnam was almost as hard on GIs as it was on civilians. The number of young Americans who died as its result is shocking.

> As our [official Army] study shook out, the fact became inescapable that a staggering 15 to 20 percent of all U.S. casualties in Vietnam were caused by friendly fire.[7]
> — Col. David H. Hackworth
> most decorated Vietnam War veteran

Of late, "surgical" air strikes, "smart" artillery shells, and pinpoint aerial surveillance have all been touted as solving the problem. Being claimed is not that the target will always be strategically important, only that its coordinates will be accurate within the satellite-ascertained Global Positioning System (GPS). Both the Chinese and North Koreans have now found ways to jam this System,[8] but the real issue is target validity. That will depend on the quality and age of the intelligence. If there are any friendlies nearby, shell load, gun idiosyncrasies, wind, and any number of other factors can still cause a deviation of up to 100 yards. This margin of error increases with map-determined coordinates, so the best way to limit friendly and noncombatant casualties is to replace standoff firepower with covert—close-range—maneuver.

Something Similar Where No Collateral Damage Occurs

The closest match to modern infantry combat is a police Special-Weapons Assault Team (SWAT) operation. There, decentralized control is a big part of what must necessarily be quick and bombardment-free. To defeat gangs or guerrillas, one has only to flood an area with enough highly opportunistic two-man roving outposts to offer each other assistance. All would have their own tiny Tactical Area of Responsibility (TAOR), with all boundaries fully known to neighboring teams. Of particular note, there are no assassination drones in police work. That's not because someone's rights might be violated, but because aerial missiles invariably result in collateral damage.

Drone strikes are seldom "surgical." Nor are they considered a legitimate way of fighting in underdeveloped regions. Their quarry may have been in the targeted structure when the intelligence was collected, but then he left. Or, his niece and her 13 children just dropped by to see him. If that quarry must be eliminated, a pair of special operators should just sneak up and shoot him. That's how the Selous Scouts used to operate.[9]

Public opinion plays a tremendous role in 4th Generation Warfare (4GW). The West no longer has the luxury of totally destroying a place to more easily liberate it. That not only antagonizes its population, but often fans the violence. Lower-intensity combat must instead be applied to the site's criminal element. Are all the residents of a U.S. inner city to blame for what a local street gang does?

To help defeat the Malayan Communists of the early 1950's, the British dispatched indigenous mantrackers after every terrorist incident. When those mantrackers caught up with the perpetrators,[10] only the bad guys got hurt.

The New U.S. Military Focus

Whoever occupies contested ground ends up winning the war, not the one flying remotely overhead. It's unrealistic to think that the U.S. can win any part of World War III (WWIII) by repeating its strategy in Serbia or Libya. It must instead have a ground presence in every contested region. Yet, that presence need not be a massive expeditionary force, only tiny increments of indigenous force multipliers. That takes semi-autonomous infantry squads to augment local police and militia detachments. (See Figure 3.1.) There aren't enough U.S. commandos to go around. Both GI types would have to become more adept at the following: (1) how to train others; (2) light-infantry subjects; (3) law enforcement; and (4) Unconventional Warfare (UW). The last will permit them to escape—by hiding or exfiltration—any encirclement.

Ever noticed how the U.S. Army has no more "straight-leg" infantry capability in its Airborne or Mountain Divisions? Dismounted grunts require more discretionary leeway than other types of troops, so the Pentagon may now deem them a liability. That would explain why all remaining (Army and Marine) infantry units ride around in trucks and fight almost exclusively with supporting arms. That approach to war only works in open terrain. In heavily foliated, precipitous, or built-up areas, there are no roads and or safe ways to bombard a nearby target. That's why it takes truly light infantrymen to influence much of the planet. Only the technology-and-rank-dependent West has chosen to ignore this rather elementary truth. Every North Korean special operator must first become a light-infantry expert.[11] (See Figure 3.2.) In other words, he knows how to get things done—in all types of terrain—through surprise and minimal force.

Additionally, all Asian Communist commandos have, as their secondary mission, to share this expertise with line infantrymen.

Figure 3.1: Patrolling against Local Disturbances of the Peace
(Source: "Martyrs' Market," by Larry Selman, poss. ©, from the following url. http://www.nationalguard.mil/resources/photo_gallery/heritage/hires/MartyrsMarket.jpg)

Figure 3.2: Light-Infantry Legacy of North Korean Commandos
(Source: Courtesy of Cassell PLC, from "World Army Uniforms since 1939," © 1975, 1980, 1981, 1983 by Blandford Press Ltd., Part II, Plate 79)

That's how North Korea's commando headquarters got is name—
"The Light-Infantry Training Guidance Bureau." Such armies will
more easily control vast regions in the next global conflict. It was
by leaving tiny detachments behind in each village while pushing
Pol Pot's forces up against the Thai border that the North Vietnamese Army (NVA) probably broke China's hold over Cambodia
in 1979.[12] Each detachment's job would have been to replace the
local Khmer Rouge cadre.

4 Enhancement of Local Security More Vital

● To what are civilians subjected in an insurgency?

● Are local police and militia often their biggest problem?

INDEPENDENCE

No country truly liberated without unimpeded local voting.

(Source: http://search.usa.gov public-domain image from this url: http://www.cr.nps.gov/history/online_books/hh/17/images/hh17k1.jpg)

Safety at the Polls a Prerequisite of True Democracy

U.S. troops are normally sent abroad to dissolve a dictatorship or shore up a republic. Who next gets elected in that country will depend on what happens at all voting sites. Thus, local security can be said to be more important than central-regime assistance. Just as the British-supervised elections of 1980 went awry in Rhodesia, so can those of any current nation under Communist or Islamist duress.

Some 7,000 guerrillas had gone to ground in the tribal

areas [of Rhodesia] on the orders of ZANLA [Zimbabwe African National Liberation Army] commander, Rex Nhongo. He told them to ignore the cease-fire, hide their weapons, and persuade the people to vote for ZANU P/F (Zimbabwe African National Union Patriotic Front) (Verrier, *The Road to Zimbabwe,* 86). . . .

When ZANLA first infiltrated its political commissars into the Rhodesian tribal areas in 1972, they embraced the . . . methods of the [Communist] Orient to politicize the tribesmen. On entering villages, they selected people for execution. . . . Their objective was to rid communities of their leadership and destroy the *bourgeois.*

Executions were conducted in an exemplary [local] fashion. . . . ZANLA's reign of terror has had few parallels in recent African history.

Not surprisingly, it took little persuasion to get [the local] villagers to set up informer networks, report on the activities of the security forces, feed and look after incoming guerrillas, and spy and report on [unseemly] reports of their fellows. . . .

After Lancaster House [the cease-fire conference], . . . [ZANLA] guerrillas were able to live openly amongst the villagers without fear of attack by the Security Forces. . . . [T]he unsophisticated tribesmen took this as a signal the guerrillas had "won." . . .

So when ZANLA guerrillas insisted that ZANU P/F would win the independence election and announced that their first task afterwards would be to open the ballot boxes and identify those who had voted against them with the help of a special machine imported from Romania, the villagers believed them. They also accepted that those who voted against ZANU P/F would afterwards be put up against walls by ZANLA and shot (Sutton-Price, *Zimbabwe,* 63).

. . . Britain's quaint idea of stationing British bobbies in uniform at polling stations to ensure fair play, was at the least naive. . . .

An admission that ZANU P/F used its notorious political commissars to intimidate the indigenous people into voting for them came during Granada Television's *End of Empire* series when Edison Zvobgo said: ". . . [W]e had a very large

army left (outside the assembly points), who remained as
political commissars . . . simply to ensure that we would win
the election (Flower, *Serving Secretly,* 255-256)." . . .
So much for the democratic process, African style.[1]
—Peter Stiff, renown South African historian

Shoring Up the Rural Infrastructure

If non-uniformed separatists do not qualify for Geneva Convention protection, then they must be afforded all the considerations of
an International Criminal Police Organization (INTERPOL) suspect.
Within a countryside that is often controlled by rebels, most local
security comes from paramilitary-police outposts. If augmented
by tiny U.S. contingents, those outposts could better prevent voting irregularities. While America's war planners almost certainly
realize this, they often consider such an assignment too dangerous
for their best special operators. They have yet to realize what a
little light-infantry, UW, and law enforcement training could do
for American commando and line infantry capabilities. Only just
recently have such planners agreed to a long-overdue shift in U.S.
counterinsurgency strategy.

The Pentagon's Latest Counterinsurgency Manual

In 2006, Generals Petraeus, Mattis, and Amos produced a joint
Army/Marine Corps manual that effectively altered U.S. counterinsurgency (COIN) doctrine. Among its most dramatic changes was
to protect a citizenry from insurgents in their midst, rather than
punish all members for allowing the presence. In COIN lingo, it
better handled the "hearts and minds" of an already distressed
population.

The aim of the this type of COIN was to win the support
of the people by protecting them, improving services, and
providing them good governance.[2]
— *Armed Forces Journal,* January/February 2013

As has been routinely the case for some of America's greatest leaders, the new strategy still tried to attack the problem from the "top down." Through supposed American expertise in countering high-level corruption, the host country's central government was first to be mended. In essence, the plan too little compensated for the "bottom-up" dynamics of an Asian society. As such, it stood little chance of removing local corruption from Afghanistan—a loosely controlled nation that depends for much of its income on a thriving drug trade.

During the Vietnam War, there had been hints on how better to pursue an anti-corruption strategy. They had come from two highly experienced Marine generals—"Brute" Krulak and Lewis Walt. Sadly, political disfavor at home was to stifle their joint insight.

American Squads Resisted Corruption in Southeast Asia

According to a *Life Magazine* article from the period, U.S. components of Combined Action Platoons (CAPs) worked against local "graft" in South Vietnam—with its inevitable links to the central government.[3] Generals Walt and Krulak had been the "godfathers" of the CAP Program.[4] Still considered—by most counterinsurgency experts—to be the only way to win the Vietnam War, it effectively turned the Maoist method in upon itself. One of its hidden benefits was a way to produce the first truly light infantrymen the Pentagon had seen since Carlson's Raiders.

Afghan Trend Was to Help Local Police and Militia

Gen. McChrystal then offered some of the same brilliance to Afghanistan that Gen. Petraeus had to Iraq. While still too centrally oriented, his new strategy showed more of a bottom-up approach to things than the new manual had allowed. McChrystal wanted widely to intermingle U.S. troops with Afghan civilians. (See Figure 4.1.) The last invader to successfully occupy Afghanistan—Tamerlane—had done that too.[5]

Of course, Gen. McChrystal also embraced the new manual's strategy of protecting the Afghan population from its criminal element, rather than treating every citizen as if he (or she) were a willing accomplice. Through similar culture, religion, or tribal affiliation, many might have previously been treated as enemy "sympathizers." This now dated perception does little to hold down the collateral damage.

Gen. Stanley McChrystal's long-awaited reassessment of the war against Taliban insurgents aims for a transformation of the shaky relationship between U.S. forces and Afghan civilians. . . .

The latest draft of the assessment also urges speeding up the training of Afghan soldiers and police and nearly doubling their numbers to roughly 400,000. . . .

The main recommendations for change stem from the military's new counterinsurgency strategy in Afghanistan, which is now designed to focus less on going after Taliban

Figure 4.1: Tribal Villages the Key to Controlling Afghanistan
(Source: "The Hizara Province," by SFC Elzie Golden, U.S. Army Center for Military History, from this url: http://www.history.army.mil/art/Golden/Image-12.jpg)

strongholds and more on protecting the local population.[6]
— *Philadelphia Inquirer,* 1 August 2009

His new strategy would also develop a central government that most citizens deemed friendly.

[T]he latest draft of McChrystal's assessment on the war includes the following recommendations:
Using intelligence less to hunt insurgents and more to understand local, tribal, and social power structures in the areas where they operate. McChrystal is considering concentrating troops around populated areas rather than going after sparsely populated mountain areas where Taliban hide.
Getting troops more active in fighting corruption. U.S. forces will need to take care in their dealings with local Afghan leaders to ensure that they are not perceived by the Afghan population to be empowering corrupt officials.[7]
— *Philadelphia Inquirer,* 1 August 2009

A Localized Approach Is Evolving Elsewhere

As readily apparent from America's ongoing Africa strategy, most efforts to resist Islamic or Communist expansion worldwide may soon revolve around tiny military contingents. The limited numbers of GIs would only act as force multipliers for indigenous security personnel. This has largely been the role of U.S. Marine special operators in Africa. (See Figure 4.2.)

In the future, smaller commitments arranged around trainers, advisers, and the use of commandos—COIN Lite—may well make the best military and political sense. In most cases, the weight of operations at the low end of the conflict spectrum is likely to fall on our hard-pressed special operations forces. . . . Regular forces [U.S. grunts] can form "advise and assist" units or *ad hoc* military assistance teams to take some of the pressure off of elite forces.[8]
— *Armed Forces Journal,* January/February 2013

To permit this same "force-multiplier "approach at higher risk

Figure 4.2: Marine Addition to Special-Operations Community
(Source: Courtesy of Orion Books, from "Uniforms of Elite Forces," © 1982 by Blandford Press Ltd., Plate 6, No. 16)

locations, the members of each tiny contingent must first become more self-regulating and familiar with Escape and Evasion (E&E) techniques. (A few of *ninpo* caliber have been included in Table 2.1).

Once a contested region has been blanketed with such teams, America's senior military officers will have less chance at a "combat command," but the collateral damage and financial outlay will be less as well.

For each isolated contingent, there is an inescapable correlation between its ability to operate semi-autonomously and to regulate its own behavior. Less dependence on any headquarters (or supporting-arms umbrella) will take more field skill than the average U.S. special-operations squad currently has. This shortfall and its other ramifications will be the initial focus of the next part.

Part Three

America Needs Self-Regulating Squads

He who knows how to use both big and small forces
will be victorious. — Sun Tzu

(Source: Sun Tzu, "The Art of War," trans. and intro. by former Marine Raider Samuel E. Griffith [New York: Oxford Univ. Press, 1963], p. 62)

5 The Pentagon's New Worldwide Strategy

- How does Washington plan to limit Islamist expansion?

- Where in the world is this new defense plan most evident?

"High-tech" U.S. troops training ill-equipped foreign forces?

(Source: http://search.usa.gov public-domain image from this url: http://usarmy.vo.llnwd.net/e2/-images/2006/11/21/990/size0-army.mil-2006-11-21-102805.jpg)

America Can't Afford Any More Massive Deployments

On 12 January 2010, the U.S. Commander in Chief announced his "post-Surge" strategy for Afghanistan. The now reinforced U.S. contingent would become a security force. Its goal would be to break the Taliban's hold on the incumbent regime, and then secure *key population centers*.[1] Within this last clause lay the same mistake the Soviets had made—forfeiting all but a few tiny pockets in the overall land mass to the opposition.

This new White House strategy hadn't been formulated in a vacuum; the Joint Chiefs of Staff must have agreed to it. Since the

Figure 5.1: Afghan Supply Conduits Too Hard to Protect
(Source: http://search.usa.gov public-domain image from this url. http://www.dia.mil/images/history/military-art/1980s-series2/field_laser_9.jpg)

Iraq pullout, American voters had been displaying less tolerance for wartime casualties. That and Afghanistan's difficult-to-protect road system must have made the outposting of its sparsely populated areas politically untenable. (See Figure 5.1.)

Upon finding the Afghan "outback" too risky to actively contest, the Pentagon then came up with a less direct way of stemming Islamist and Communist expansion.

Backed by a new policy geared toward quelling African-based terror groups, the Pentagon is going on the offensive on the continent, setting up what could be the template for the next-generation of U.S.-led counterterrorism operations worldwide.

The approach that U.S. counterterrorism forces take in Africa will likely be less defined by night raids and other direct action missions that dominated operations in Iraq and Afghanistan.

Rather, American special operations troops and supporting forces will be focused on indirect missions, characterized by cooperative efforts in military training and logistics support to partner nations in Africa.[2]
— *The Hill,* 14 July 2012

America's foreign-assistance emphasis had shifted back to the old standby—training and logistical support. Had the Pentagon learned nothing from the last days of South Vietnam or initial attempt to capture Bin Laden? It was indigenous forces that had let the *al-Qaeda* leader through the cordon at Tora Bora. For the Pentagon's new counterterrorism plan to work, U.S. troops would have to be more deeply involved than that. Even maneuver element advisers had been unable to stop South Vietnam's eventual slide into Communism.

The Telltale Kony Failure

The future prospects of this new global strategy are best assessed from the progress in Africa against the Lord's Resistance Army (LRA). The leader of that criminal militia—Joseph Kony—is now no closer to being stopped than when President Obama first ordered it in November 2011.

U.S. military commanders said Sunday that they have been unable to pick up his trail but believe he is . . . hiding in . . . dense jungle, relying on Stone Age tactics to dodge his pursuers' high-tech surveillance tools.[3]
— *Washington Post,* 29 April 2012

The problem is clear. The U.S. special operators assigned to the task had been relying on local forces to provide enough human intelligence and mantracking skill to get Kony. That's not how the world-famous Selous Scouts would have handled such a mission.[4] But then, the average Selous Scout had more field skills

Figure 5.2: More Field Skills Necessary in Heavy Vegetation
(Source: "Indiana Rangers," by Mort Kuntsler, poss. ©, from these two urls: http://www.history.army.mil/art/225/INRGR.jpg and
http://www.nationalguard.mil/resources/photo_gallery/heritage/hires/Indiana_Rangers.jpg)

(and delegated authority) than his American counterpart. In any jungle engagement, "woods smarts" are often the deciding factor. (See Figure 5.2.) Kony had become too crafty to leave an electronic trail, and the U.S. contingent lacked the field expertise or political leeway to close with him any other way.

LRA fighters have slipped deeper into the bush, splintering into smaller bands to avoid detection and literally covering their tracks. . . .

. . . [H]is 200 or so fighters rely on foot messengers and preordained rendezvous points to communicate.

Kony's methods have proven effective against the U.S. military's satellites . . . and other forms of [electronic] surveillance. Commanders warn that it could take years to find him.[5]

— *Washington Post*, 29 April 2012

While those local mantrackers were probably good enough to locate Kony's *entourage,* their assault backup may have been lacking. Such problems can run the gamut from "a payoff to let him escape" to "not knowing enough about maneuver to take such a force by surprise." Any U.S. helicopter insert would have doomed the attempt. Until all U.S. special operators are light-infantry experts, that level of training assistance will not be possible for indigenous forces. If a few U.S. "observers" had accompanied the search force, at least the disconnect would have surfaced. But, no GI had left camp.

> Since October, U.S. troops have fanned out to five outposts in four countries, advising thousands of troops from Uganda, the Central African Republic, South Sudan, and Congo. . . .
> The Americans said they rarely leave the vicinity of their camp. . . .
> The U.S. forces carry arms but are not permitted to engage in combat, except in self-defense.[6]
> — *Washington Post,* 29 April 2012

Only Part of Problem Has Been Solved

Sensing too little progress with the old "equipment and training on how to use it" routine, the American President may have directed more U.S. participation in host-country ground operations.

> The U.S. government declared the LRA a terrorist organization a decade ago and has provided the Ugandan military with equipment and advice for years.
> But Obama raised the stakes of American involvement in October by ordering troops into the field.[7]
> — *Washington Post,* 29 April 2012

With more light-infantry skill, America's special operators could have better capitalized on this break in long-standing dogma. To make any real difference, they would have to do more than just train indigenous forces. And to assist every African country under duress, they would need the help of scores of specially prepared U.S. infantry squads.

Pentagon Plan Still Has Possibilities

As the world continues to change, so must the Pentagon's defense strategy. In truth, the above way of curbing Communist and Islamist expansion is "spot on," with one or two qualifications. As shown by this chapter, any U.S. military aid must entail an active presence at the local level to prevent voting irregularities and a repeat of the Kony fiasco. That's hard to do safely with anything less than a squad-sized unit. And that unit, whether of special operators or line infantrymen, must be more tactically adept than at present.

By some estimates, the number of Mainland Chinese special operators has now risen to one million of what used to be a 1.6-million-man ground force. In 2009, the North Koreans also upped their commando count from 100,000 to 180,000.[8] That's how the Communists plan to bypass Western firepower in their ongoing quest for political and economic domination of the world. Only this time, it will be "death by a thousand razor cuts" on a much larger scale and then a sweep of leftist-election victories. The U.S. doesn't need an equal number of commandos to counter this offensive, only to make better use of its line infantrymen.

Many of the Chinese peacekeepers and corporate guards have been protecting their places of foreign assignment from "pro-democracy" activists. Yet, such meddling has been far from defensive in nature. Under the threat of further U.S. intervention in Haiti during February 2004, the PRC dispatched a "special police contingent" to a tiny and highly distant nation that still diplomatically recognized Taiwan.[9] Those who see only goodwill and coincidence have never read Sun Tzu.

Again evident around Aleppo in Syria's civil war,[10] foreign assistance from Lebanese *Hezbollah* or the Iranian Revolutionary Guard also takes the form of a single squad per neighborhood.[11] This is the Asian (bottom-up) way of engaging a target. To counter it, the Pentagon must become more involved at the local level. Instead of just advising or training forces that may occasionally pass through each contested settlement, a squad of GIs must be actually stationed there as part of its paramilitary police contingent. By living off the local economy, those tiny groups of Americans would require little, if any, logistical support.

How Best to Train Local Security Forces

- How do foreign paramilitary units operate?
- Which GIs are best qualified to train them?

With enough field knowledge, locals don't need much equipment.

(Source: http://search.usa.gov public-domain image from this url: http://www.history.army.mil/images/artphoto/pripos/wwii-tideturns/BackRoad.jpg)

All Cultures Have Combat Strengths

Most "developing nations" lag far behind the West in military technology and firepower. Yet, their residents still know something about self-defense. If those nations are heavily wooded or rurally populated, many of their military-aged males have more "woods smarts" than the average GI. With the right foreign instruction, this natural edge could easily be brokered into an advanced light-infantry (special-operations) capability.

Almost by definition, "light" infantrymen can compensate for any lack of firepower with more surprise-oriented maneuver. For the young African, secretly approaching an enemy sentry is little different from sneaking up on a deer. All he has to be shown is how covertly to bypass (or silence) that sentry. Then, he and 12 of his buddies are well on their way to state-of-the-art (post-machinegun) squad assault technique. While their target-acquisition capabilities may not be enhanced by man-packed electronics, neither is their bullet-dodging agility degraded by extra equipment. Thus, they have just as good (or possibly even better) odds of avoiding personal injury during a contested assault. After all, night-vision goggles, laser-projecting muzzles, and bulletproof vests do nothing to reduce an attacker's visual, auditory, olfactory, or other signatures. Those can only be minimized through blackened skin, lack of gear/clothing, careful diet, inclement weather, and microterrain utilization.

Who Makes the Best Foreign Instructor

For a nonresident to succeed as a force multiplier or trainer of "Third World" militia, understanding the local culture must come first, then how "militarily" to capitalize on that culture. (See Figure 6.1.) Any population with a hunting (albeit poaching) heritage is virtually awash with animal trackers. The basics for interpreting/following the trails of men and animals are not that different. Where governmental anti-poaching efforts are in evidence, skills for following men will automatically exist. They are also present along heavily contested borders. Within the mountainous regions of India (along its Pakistani and Chinese frontiers), reside "Khojis." They are border police mantrackers similar to those in Israel.[1] Their human-sign-reading abilities could easily be expanded into world-class offensive and defensive maneuvers, such as short-range infiltration and roving listening posts.

Advanced assault technique normally calls for some occasion of reduced visibility. Almost all East Asian armies have a night-fighting tradition that involves enemy fortress penetration by tiny, "extraordinary" elements. Isn't that what Japanese *ninjas* (those who sneak in) and their elsewhere-copied Chinese predecessors—the *moshuh nanren* (gap men)[2]—used to do? Since 1941, many lower-ranking U.S. service personnel have had painful experience with this more obscure part of almost every Japanese,[3] Chinese,[4] North Korean,[5] and North Vietnamese attack.[6] All the while, their com-

Figure 6.1: Other Societies Have Own Ways of Fighting
(Source: "Chat among Friends," by SFC Darrold Peters, from U.S. Army "War on Terror Images," at this url: http://www.history.army.mil/books/wot_artwork/images/36b.jpg)

manding officers and manual writers have been mostly recalling the concurrent feint—often a human-wave "demonstration." Not surprisingly, all four cultures also like to tunnel (fight in the least obvious of the three dimensions).[7]

Other South and Southwest Asian societies—like the Iranians[8]—have a local heritage of widely dispersed, semi-autonomous elements operating against a more powerful foe. Behind it are often religious roots. Islam's breakaway Shiite sect first gained momentum after Husain (one of the 4th caliph Ali's sons) was killed leading 72 men against 400 at Karbala.[9] Through *al-Qaeda's* training liaison with Lebanese *Hezbollah,[10]* this Shiite propensity toward martyrdom has now reached many fundamentalist Sunni factions. Few Muslim men, of whichever persuasion, would thus object to operating—while severely outnumbered—in the small-group size that has come to define modern war. (See Figure 6.2.)

51

Figure 6.2: Many Muslims Like Working Alone
(Source: "Masters of Chaos: ODA 563, 5SFG(A), Dec. 2001," by MSG Christopher Thiel, from this url: http://www.history.army.mil/art/Thiel/Images/AI%20inODA563.jpg)

What Not to Teach

Other societies can be differently motivated in battle. For example, most Afghan fighters prefer vociferous displays of *bravado* to any covert activity. The Pakistani head of anti-Russian resistance soon realized this.

> [W]hat was wrong with my method was that it lacked noise and excitement. It was not their [the *mujahideen's*] way to

fight, with no firing, no chance of inflicting casualties, no opportunity for personal glory and no booty.[11]
— Brigadier Yousaf, Afghan Service Bureau Chief

Afghanistan is a very poor country. Its tribal way of settling differences is openly to bombard one's quarry into submission, and then take all his booty. This seldom involves an on-line assault. It more often resembles a gradually closing cordon that will only infringe upon the quarry's perimeter in daylight. That penetration is accomplished a little at a time, and at multiple points if possible, just as with a Western "fire and movement" attack. Like NVA soldiers, Afghan tribesmen do not consider closely encircling a foe to be fratricide.[12]

Their [the *mujahideen's*] method was to bombard the [enemy] posts with heavy weapons by night at long range, move closer to fire mortars, get 30-40 men to surround them, and at short range [presumably at dawn for better target acquisition] open up with machineguns, RPGs and RLs [rocket-propelled grenades and rocket launchers].[13]
— Brigadier Yousaf, Afghan Service Bureau Chief

Because of the intrinsic value of what stands to be captured, nothing is done during the attack that might endanger its on-hand quantity. That its existence makes the position more defensible is never even considered.

If the garrison withdrew, the posts were captured and the *mujahideen* secured their loot in the form of rations, arms and ammunition, all of which could be used or sold. Then, only then, was the charge laid on the fuel pipeline. If the garrison stuck it out, the pipeline remained untouched.[14]
— Brigadier Yousaf, Afghan Service Bureau Chief

Seldom does this type of activity lead to the crawling or digging of advanced assault technique. As a result, this region's tribesmen and drug runners have never discovered (or reinvented) any of the Stormtrooper assault variations.

Mujahid hated digging. . . . [A] static defensive role . . . was alien to his [commander's] temperament; it restricted his

53

Figure 6.3: Martyrdom More Likely through Direct Confrontation
(Source: "Hill 609 [Tunisia]," by Fletcher Martin, U.S. Army Center for Military History, posted Oct. 2005, from this url: http://www.history.army.mil/art/A&I/Hill_609.jpg)

freedom to move, and he could seldom be convinced of the need to construct overhead cover. Similarly, his fieldcraft was often poor as he was disinclined to crawl, even when close to an enemy position.[15]

— Brigadier Yousaf, Afghan Service Bureau Chief

Other Influences on the Curriculum

Afghan fighters have almost no tradition of a noncommissioned officer (NCO) corps. They have only experienced the more direct officer (or tribal-chieftain) to fighter relationship.[16] Thus, any number of Asian counterparts may share their lack of tolerance for personal criticism. Trying too quickly to rid such a group of "bad habits" might not only limit individual growth, but also undermine unit cohesion.

Before any such catharsis is attempted, the Western instructor must be doubly sure that his own squad tactics are highly enough evolved to generate almost total surprise (allow an undetected approach to within 10 yards of defenders). East Asians will be unimpressed by anything less. Muslim residents of South Asia, the Middle East, and Africa may settle for a long stand-up assault, but only because it does so little to forestall their martyrdom. (See Figure 6.3.)

Having mostly relied on fire superiority, Western instructors would be better off building upon local strengths than discouraging local weaknesses. The U.S. military, because of its relative wealth, has never put much emphasis on squad maneuver. This may have skewed its priorities a little. For one thing, short-range combat does not require as much accuracy in shooting. Nor can it be mastered by strictly obeying all instructions.

Though a Third World resident's lack of discipline may at first seem a hindrance, it may also signal enough self-confidence to generate a little personal initiative. Such a thing becomes more important in Asia, where everyone tends to follow the crowd. Any Western unit that becomes too highly structured or controlled may find itself deficient in this area as well. More important than unit discipline in a highly dispersed and chaotic environment is self-discipline. That is one of Evans Carlson's greatest legacies. He, after all, had been instructing his WWII Raiders on how best to utilize indigenous personnel. This insight may have come from Mao Tse Tung himself.

[T]he basis for guerrilla discipline must be individual conscience. With guerrillas, a discipline of compulsion is ineffective.[17]
— Mao Tse-Tung quote
from 1941 *Marine Corps Gazette*

While trying to remake such a group in one's own image is risky, many may still retain the mannerisms of a past occupier or adviser.

Prospective Soldiers from Formerly Soviet Countries

Citizens of former Soviet "republics" and client states may display unique characteristics. The former republics are along the ancient Silk Route and in Eastern Europe. A few of the client

Figure 6.4: Selous Scout NCO
(Source: Courtesy of Orion Books, from "World Army Uniforms since 1939," © 1975, 1980, 1981, 1983 by Blandford Press Ltd., Part II, Plate 98)

states are Egypt, Syria, Cuba, Angola, and Venezuela. Possibly due to Russia's Far Eastern heritage, its troops became quite good at night fighting during the latter stages of WWII.[18] A U.S. Army study estimates that 40 percent of all Soviet attacks in 1944-45 were at night.[19] Even the tactically advanced Germans have confirmed this.

> The Germans . . . admitted that their [Russian] opponents were better than they at fighting at night, in forest and swamps, at camouflage, and quick digging in.[20]
> — Republisher's introduction
> *Soviet Combat Regulations of November 1942*

Most senior German officers who fought the Soviets on the Eastern Front acknowledged their [the Soviets'] "natural superiority in fighting during night, fog, rain or snow," (Simon, in *Night Combat*, by Kesselring, 1952, 59) and especially their skill in night infiltration tactics, reconnaissance, and troop movements and concentrations (Kesselring, *Night Combat*, 1952, 2-5).[21]

— U.S. Army, *Leavenworth Papers No. 6*

When in Afghanistan during the late 1980's, Soviet Army units routinely established ambush and blocking positions before dawn.[22] Their zeal for completing a mission under cover of darkness had started to wane.[23] Still, the residents of any Soviet affiliate may like fighting at night. This would make it far easier for them to embrace advanced assault technique.

Nor were Soviet soldiers the least bit hesitant about crawling in heavy combat.[24] Even during a nighttime assault, WWII Russians regularly hit the dirt every time they were illuminated.[25] Westerners tend to move up streets and hallways in urban combat, but Russians have traditionally avoided such kill zones—preferring instead to move through yards, gardens, and holes in walls.[26] Any developing-nation citizen who had watched them train would be more likely to follow that example.

Some of the World's Most Promising Fighters

Any participants of a Maoist insurrection anywhere within Southeast Asia, Africa, or the Western Hemisphere would thrive on advanced light-infantry instruction. Though that former rebel might pose something of a security risk, he still has more potential than his average countryman. That's because his field skills and initiative have both been recently exercised.

While facing indigenous Maoist guerrillas in Southern Africa, the world famous Selous Scouts routinely jumped at the chance to turn a captive into a trusted unit member.[27] (See Figure 6.4.) Though the threat of execution may have helped with the initial conversion process, the emphasis on small-unit action (and its prerequisite delegation of authority) is what led to the enduring union. As naturals for any light-infantry mission, these former guerrillas became highly regarded for more than just their intelligence value.

How Is Advanced Assault Technique Tied to Light Infantry

"Advanced" assault technique is that which allows the least number of friendly casualties. If so much as one defender survives a shooting assault, he can inflict serious injury upon the main body of attackers. That's why the safest way to seize an objective is so totally to surprise its defenders "up close" that none can resist. Light infantrymen carry less ammunition to increase their battlefield agility. As such, they become much more proficient at getting this near undetected.

Light Infantrymen Need
_____ No Technology

- Why are "light infantrymen" called that?
- Do developing nations prefer them to the U.S. kind?

Enough surprise in an assault can take the place of firepower.

(Source: http://search.usa.gov public-domain image from this url: http://www.history.army.mil/images/artphoto/pripos/amsoldier/5/1945_Philippine.jpg)

A Recurring Claim

Though it has been often said that the NVA needed no tanks, artillery, or airstrikes to evict U.S. forces from Vietnam, most modern-day GIs don't believe it. They think their fathers and uncles won every ground engagement and were then sold out by the *New York Times* or Congress. In truth, the enemy did use a few PT-76 light amphibious tanks, some long-range artillery from the Demilitarized Zone (DMZ), and possibly even a helicopter pickup or two near the Laotian border. But, compared with the Allies' constant reliance on supporting arms in Vietnam, its opposition needed no technological "crutch" to do what it wanted. That's how truly "light" infantrymen operate. They can generate so much surprise tactically as to require little firepower on either offense or defense. For them, any outgoing mortar or artillery shell is normally a feint—a noise-masking ruse to lead a firepower-oriented quarry to believe there are no enemy assault troops nearby.

What has lately been called "light infantry" in the U.S. inventory is really of the heavy or "line" variety. (See Figures 7.1 and 7.2.) It fights with supporting arms and travels by truck.

Figure 7.1: How Army Remembers "Tracking Bin Laden" in 2002
(Source: "Tracking Bin Laden," by SFC Elzie Golden, U.S. Army Center of Military History, from this url: http://www.history.army.mil/art/Golden/Tracking.jpg)

Figure 7.2: "Heavy" Grunts Little More Than CO Extensions
(**Source:** U.S. Air Force Clipart Library (www.usafns.com/art.shtml), image designator "1-4e.tif")

How U.S. Ground Forces Operated during Vietnam War

While countering guerrillas in Vietnam, U.S. rifle companies seldom just peacefully patrolled the countryside. After much of the well-populated coastal plain had been declared a "free-fire zone," they repeatedly swept it in full battle array. Any incoming fire was met with overwhelming bombardment. That's how the "pacification" was conducted. Any antagonist not willing to stay and fight was considered a coward and (depending on his attire) unworthy of Geneva Convention rights.[1] Such is the folly of trying to solve a 4GW problem with 2nd Generation Warfare (2GW) methods. If one's opposition decides to operate piecemeal, then each of his tiny elements will have to be handled with minimal force. Otherwise, too many innocent bystanders get in the way.

A Strange Lack of Respect at "Three Gateways to Hell"

There was a place midway between the DMZ-adjacent bases of Gio Linh and Con Thien that the Marines called "Three Gateways to Hell." It was here that the Westward-leading trail topped a low rise, which was later proven to hold a major North-South infiltration route.[2] Just to the right of this trail in March of 1967, a Marine Lieutenant saw two fully exposed NVA leaders watching a nearby firefight unfold. Only the Lieutenant could see the pair, because seven-foot high bushes separated them from the rest of his trail-bound platoon. Only later, did he realize why those enemy leaders had been so nonchalant.

The Asian Communist officer is known to prepare his unit differently from an American. Instead of always teaching his men something, he often just watches to ascertain their abilities. Then, in combat, he tries to arrange for commensurate circumstances.[3] That might explain some of what the U.S. Lieutenant had already witnessed over seven months in country. But, at Three Gateways to Hell, there had been a complete lack of concern for what U.S. forces could do. This so aggravated the Lieutenant that—though fully visible to a nearby enemy machinegun—he started to fire at the two NVA leaders with his pistol. What happened next would be laughable if not so nearly fatal. After missing them so badly with his first six rounds that they didn't even flinch, he did what seemed like the fastest reload in Marine Corps history and then tried to draw a better bead. When his men later found him, there was no round in the chamber of his elsewhere-located handgun. After inserting the new magazine, the platoon leader must have failed to pull the pistol slide hard enough to the rear.[4]

Only many years later did the full meaning of this encounter sink in with the U.S. officer involved. The leaders of an NVA platoon had so little regarded its American counterpart, that they felt perfectly comfortable just standing around as the two made contact. Whether their opinion was, in any way, justified will take further discussion.

All Details Surrounding the Event

At the center of this story was a platoon from A company, 1st Battalion, 4th Marines. It had been moving just ahead of the bulldozers clearing the new McNamara Line. Right before the rise in the trail, its lead element had run into a wide expanse of tall bramble bushes along its north side. That's when the platoon leader decided to honor his right-hand flankers' request to rejoin the column. Lt. Jack "The Georgia Peach" Cox had been killed within 150 yards of this very spot while trying to outflank a battalion with a platoon, so all buddy teams would have to stay within easily supportable distance.

It wasn't long before a burst of machinegun fire shattered the morning calm. The Lieutenant's hunch had been right. The somewhat sunken path had been so heavily lined with bushes that no trail occupant could be specifically targeted from the side. That's why he and all his men were still standing. There were a few freshly dug fighting holes beneath the closest foliage, but all were empty. The Lieutenant pushed through a tiny opening in the hedgerow to where he could see a small, pockmarked clearing. He called for a machinegun team. Shortly thereafter, the lead squad leader came over wanting a mission. After the machinegun had arrived and was starting to return fire, the Lieutenant said to the others nearby, "Follow me."[5]

That no one did so should come as no surprise or be construed as any reflection on their professionalism. Moving directly into automatic weapons fire across open terrain is not particularly smart, even where shell holes exist. Unfortunately, the Lieutenant was already committed. He vaguely remembers zigzagging past the first crater to give those behind him a place to take cover. After running a few more yards, he couldn't help noticing two things: (1) there were no more shell holes in that direction; and (2) just to the left and 30 yards ahead were two bareheaded enemy soldiers just standing there. They weren't looking at him, but over to where their men must have been. The young Marine then deduced from his training that the only honorable thing left to do was to try to shoot them.[6] So, he took a knee and rapidly expended all six rounds in his pistol. After no visible reaction from the two, he quickly replaced the magazine and was about to "hold and squeeze" another shot when everything went black.

Only out for a few seconds, the Lieutenant noticed a tiny, oil-stained hole in the groin area of his trousers. He then remembered the last shell hole. As he crawled the 10 or so yards back to it, many little "bees" started to buzz past his head. One must have grazed his ankle as he dived into the crater (or the first bullet had hit him twice). Next, he looked up to see no fewer that five "chicom" grenades coming his way. He quickly curled up, but the closest was to explode at the lip of his hole. After so near a miss, the Lieutenant never even considered tossing a grenade of his own. Within a few minutes, ten or so Marine mortar rounds rained down on where the enemy had been, and a Corpsman showed up. He announced that the officer's leg was not broken. After just crawling ten yards on a shattered femur, his new patient begged to differ and asked to be left alone. He was then loaded aboard a medivac chopper with no splint, and one of the walking wounded kicked the leg on his way in.[7] (See Figure 7.3.)

As later recounted by the lead squad leader involved, there was a subsequent ground attack of sorts against the above-mentioned enemy position.

S.Sgt. Burke came back to the squad and said the squad was going to be [the] envelopment and assault the NVA. . . . The squad was fired up and jumped through the hedgerow and shifted to the left and move[d] West and sta[r]ted the assault. S.Sgt. Burke followed us in the assault. We were on line, and we were very fortunate in that the NVA [had] pulled back and left some token resistance. We moved forward in the assault about 80 meters and stopped. I am really glad they left that area. We found 2 dead NVA soldiers which were left behind, and we killed when they fired at the squad, and we seen *[sic]* some blood trails of the other NVA [who] pulled away. What amazed me when we assaulted through the automatic weapons position? We wanted to pursue the NVA but Staff Sergeant Burke said stop, and the squad set up in a temporary 180 [degree defense], and found out some things about our new M-16. It does not work like it was suppose[d] to. Everyone's weapon in the platoon jammed. We also notice[d] all kinds of land line wire all over the place, which was forest green in color. The wire was leading in all directions. It reminded [me] of the land lines

Figure 7.3: The Trustworthy UH-34 "Choctaw"
(Source: Courtesy of former Marine Michael Leahy, © n.d. by Michael Leahy)

we had seen in March when we were making heavy contact with the NVA [at the same location]. It is hard to believe that the two M-60 machine guns were firing as much. . . .

Captain Corcoran came to S.Sgt. Burke and said that he and my squad did an excellent job and it was [a] textbook envelopment.[8]

— Tag Guthrie, former squad leader with A/1/4

In all fairness to the platoon's other members, this young officer had been somewhat disconsolate for weeks. After questioning Sgt. Price's fondness for sneaking up on sleeping troops from outside the perimeter, the Lieutenant had been given a new Platoon Sergeant and "volunteered" for an impending transfer to Division Recon. This

Figure 7.4: Vietnam War Largely Fought from Behind Barbed Wire
(Source: http://search.usa.gov public-domain image from this url: http://www.quartermaster.army.mil/oqmg/Professional_Bulletin/Units/Images/C_225th_Fwd_Spt_Bn.jpg.")

compounded the discomfort he already felt for so much responsibility. He had never been too thrilled about making decisions that might cost 42 men their lives. Much of this disquiet had come from rather average leadership ability, uncertainty as to his own level of tactical proficiency, and too much beer at Gio Linh.[9] (See Figure 7.4.) Only 25 years later would he discover that his diminutive and little-regarded foe had been "equipped" with state-of-the-art assault technique (German Stormtrooper stuff with mortars instead of artillery, and satchel charges instead of concussion grenades).[10] At the time, he had only known that the Corps tradition was for all officers to lead by example.[11]

That was a long time ago. For those who believe short-range warfare to have significantly changed, a more modern example will be necessary.

What Truly Light Infantrymen Can Still Accomplish

As recently as 2006, a fully supported Israeli sweep into Southern Lebanon was thwarted by a few *Hezbollah* fighters below ground and elsewhere-located "medium-tech" gadgets. Those tank-killing munitions had been fired remotely. Because there were no collocated human spotters or trigger-pullers, the Israeli aircraft had no heat signatures to target.

> [An] epochal shift [in anti-armor warfare] had occurred . . . in the small village of Bint Jbeil . . . and . . . defile of Wadi Saluki, where Hezbollah fighters ambushed and destroyed a battalion's worth of Israel's blitzkrieg heavy tanks.[12]
> — Maj.Gen. R.H. Scales U.S. Army (Ret.), 2007

This same well-respected general predicts that with arms from the second precision revolution, "skilled infantrymen will make mechanized warfare a relic of the Machine Age."[13] Much of that "skill" must necessarily be in light-infantry subjects. For, only with that, will those infantrymen remain hidden and thus safe from enemy fire.

The ease with which this Islamist movement (as opposed to an organization) faced down 350-400 main battle tanks and their sophisticated air cover is troubling. (See Figure 7.5.) *Hezbollah's*

Figure 7.5: Israeli Armor Had Little Chance at Bint Jbeil
(Source: U.S. Air Force Clipart Library (www.usafns.com/art.shtml), image designator "scott_et_al_long%20war%20presen copy07 copy.jpg")

67

willingness to tunnel and disperse certainly played a role. Then, through a combination of video camera and computerized missile, it had destroyed over 50 of those tanks (by the Israelis' own count).[14] Of course, some armor had been lost the old-fashioned way. During Israel's occupation of southern Lebanon, *Hezbollah* had made a veritable art form out of remotely detonating a below-surface charge wherever Israeli tanks became canalized. This charge did not have to be "shaped" to penetrate the tank's underbelly, only big enough to lift its whole daunting mass off the ground. When the Merkava came back down (ostensibly in tact), all crew members were dead from concussion. Still, more than a good defense is possible from light-infantry expertise.

8 _ Taking Strongpoints without Bombardment

- Why must defense belts be "softened up" prior to an attack?
- Won't that alert belt occupants and risk hitting friendlies?

This helps little where bastion's front crisscrossed by MG fire from its back.

(Source: public-domain material, from U.S. Army "War on Terror Images," at this url: http://www.history.army.mil/books/wot_artwork/images/19b.jpg)

VC Needed No Bombardment to Penetrate U.S. Bases

Few American bases in Vietnam escaped sabotage from within. Though barbed-wire-and-sentry-protected equipment, fuel, and ammunition were regularly blown up, base commanders seldom realized the porosity of their lines. Besides a little misplaced pride, they suffered from two misconceptions: (1) that a damaging explosive necessarily comes from afar; and (2) that a nearly naked peasant can't personally deliver it. That's why indigenous workers were so often accused of pacing-off mortar targets. The more likely threat was from a mortar-fin-toting sapper. After sneaking into the base, he had only to plant a timed charge, drop the fin, depart the same way he entered, and then "poop" in one or two hastily aimed mortar rounds. To the firepower-oriented camp commander, all strategic-asset-destroying blasts were then nothing more than "lucky mortar hits." While mostly ignored, the same thing had been happening to U.S. forces since Guadalcanal.[1] That's because deception—and surprise in general—are more heavily valued by Oriental armies. The Asian Communists have been steadily improving a way to negate most advances in Western technology. After being unceremoniously removed from Korea following the Chosin Reservoir evolution, Lt.Gen. "Chesty" Puller alerted his fellow Americans to this disquieting fact. His precise words are as follows: "The Communists have developed a totally new kind of warfare. . . . This is total warfare, yet small in scope, and it's designed to neutralize our big . . . weapons."[2]

The key to this new way of fighting is something that only a frequent traveler to the Far East would even suspect—a higher level of skill and initiative from the average Communist soldier. This is evident from something that the architect of the Viet Minh and VC/NVA victories later let slip.

> Whether in guerrilla . . . or limited regular warfare, . . . it [armed Communist struggle] is fully capable of . . . getting the better of a modern ["high-tech"] army. . . . This is [through] the development of the . . . military art, the main content of which is *to rely chiefly on [the] man, on his patriotism and . . . [individual] spirit.* [Italics added.] [3]
>
> — Gen. Vo Nguyen Giap

The Dong Ha Experience

What is to follow is not fiction, nor is it the attempt of an inveterate bench-warmer to look like anything else. It happened just as described and actually brings into question the involved officer's level of professionalism.

The "full-service" Marine base closest to Vietnam's DMZ had been given the name of a nearby village—"Dong Ha." It had its own airstrip, artillery batteries, ammunition/fuel dumps, and sophisticated communications facility. From there, one could just barely make out the Con Thien hill mass to the northwest. Regularly supplied by shallow-draft U.S. naval craft coming up the Cua Viet River from the South China Sea, Dong Ha seemed almost impervious to attack. It even had a big South Vietnamese Army (ARVN) encampment virtually nextdoor to it. As both were on rolling, scarcely vegetated terrain reminiscent of southern Idaho, the whole area seemed completely secure.

In September 1966, A Company 1/4 was manning an eastern sector of the Dong Ha lines. Just off the runway, this string of partially fortified fighting holes had several rows of barbed wire to their front. Some 1,000 yards farther up a gentle slope was the ARVN compound. Because of the unusual terrain, line occupants could acquire a faint horizon after dark by simply lowering their heads to surface level.

For several nights at about 3:00 A.M., the members of one platoon had been waking up their Platoon Leader with reports of movement in the wire. When the recently graduated gift to warfare finally went to see for himself, he thought he saw it too. At this point in the perimeter wire just happened to be a "gate" for friendly patrols to use. So, the young Lieutenant crawled out to investigate. Fortunately for him, nothing materialized; and he returned to his "sack." Then, shortly after dawn, he went out again—this time all the way through the zigzag path through the wire. As he had no mantracking skills, this was more an act of *bravado* than anything else. After looking around for about 50 yards beyond the gate, he turned to go back to the base. Then, on the impulse of a 23-year-old, he spun around to look behind him. There, it was—something so incongruous as thoroughly to confuse him. Within 30 yards of where he stood was the gear-free caricature of an ancient *ninja* warrior running from right to left through the occasional waste-high bushes. Almost immediately, the dark figure disappeared behind a hillock no bigger than 30-feet wide and six-feet high.

Not quite sure what to make of the sighting, the Lieutenant pulled out his pistol and walked over to the very limited area in which the man had vanished. After stomping on every inch of ground and pulling on every bush in systematic search, the young officer finally returned to his regular duties without ever reporting the incident. (See Figure 8.1.) At the time, he thought the man must have been an ARVN commando from up the hill using U.S. lines as a mock-intelligence objective. After all he had heard about the absolute superiority of U.S. troops over any technologically deficient rival, what else could that figure have been? [4]

Figure 8.1: Actual Dong Ha Figure Had No Weapon or Satchel
(Source: Cover art from "Phantom Soldier," courtesy of former Marine Michael Leahy, © 2001 by Michael Leahy)

A Follow-On Sighting at the 27th Marines

At the start of a second overseas tour, this same Lieutenant served five months as Headquarters Commandant of the 27th Marines. Despite the flowery title, he was nothing more than a headquarters detachment commander who mostly oversaw the day-to-day functioning of its billeting area. He might have written off the Dong Ha sighting as unique to the war, if not now confronted by a firsthand account of something similar.

In the summer of 1968, 27th Marines had its command post (CP) well south of Da Nang and a few miles north of Hill 55. At one end of a small hamlet, this encampment sported rice paddies on three sides as its only natural protection. Other than the radios to monitor three semi-autonomous battalions, it contained nothing of particular interest to the enemy. There was no ammo, fuel, or armored vehicles. Yet, something happened one night that would help that young officer to better understand his Dong Ha experience. At about 10:00 P.M., a peer came running out of the poorly lit officers' tent area excitedly to report a surprise encounter with a black and unencumbered figure. Just as startled, the intruder had run between the "hooches," clothes-lined himself on somebody's wash, and then disappeared into the darkness. This time the Dong Ha secret keeper reported the incident to his reporting senior, the Regimental Executive Officer. Upon being informed of the incident, the 27th Marines Commander decided to do nothing about the unwelcome visitor. Somewhat surprised at the decision, the young officer still saw the wisdom in it. If someone were so skilled as to sneak through several echelons of barbed wire and sentries, any attempt to find him now would be a waste of time. Still, in this young officer's mind, the similarity between the two sightings had made a lasting impression. After a little post-retirement research into the capabilities of other nations, it all added up to only one thing—most Asian Communist commandos are *ninjutsu* trained.[5] Anyone who doubts this has only to Google the term *"dac cong."* That's Vietnamese for commando or sapper. It may also be the name of an enemy naval commando force formed in March of 1967.[6]

More Variations on the Same Theme

Besides carefully disguised sabotage, a few short-range infiltrators can do much to facilitate an all-out attack. That's how the NVA liked to use local rebels. One or two VC would fully map out an objective from the inside and then later clear a path through the obstacles for an NVA assault unit. Or, as in Korea, enough sappers would sneak into a U.S. position to destroy its command-and-control, fire support, or other central components. Normally, there was a conventional attack concurrently in progress at the position's periphery.

After sufficient training, U.S. personnel could do the same kinds of things. Without any preparatory bombardment, they could use short-range infiltration to cut the heart out of most defensive positions. For a barbed-wire-free bunker complex, the ideal would be a daytime crawling attack that required no long-range fires whatsoever (or possibility of "short rounds").

An official study of U.S. losses in Vietnam has put those from "friendly fire" at between 15 and 20 percent.[7] Most must have come from errant indirect-fire or aerial munitions. No amount of preparatory fire will kill every dug-in machinegunner anyway. (See Figure 8.2.) Plus, a tank-infantry assault involves proximity to dangerous treads. Why take either risk, when it has been widely acknowledged that enough surprise will compensate for any lack of firepower? A really proficient attacker could just stick a gun into every defender's ear and ask for his surrender. 4th Raider Battalion realized this at Bairoko on New Georgia after their air support failed to show up one steamy morning in July 1943. From then on (as noted in Chapter 3), it and its line infantry redesignation—2/4—preferred to assault strongpoint matrices without any fire support.

Sister Battalion Reaffirms Raider Paradox in Vietnam

Besides causing friendly casualties, standoff weaponry requires so much coordination as to make attack unit momentum hard to sustain. Still not totally accurate, its impacts must be kept well away from any assault element.

The 3rd Marine Raider Battalion (soon to be redesignated 3/4) often worked alongside Carlson's Maoist outfit in WWII. Company M was actually attached to 2nd Raider Battalion for a while.[8] During the Vietnam War, M/3/4 was to display the same lack of interest in artillery or air support during a heavily contested assault on mutually supporting bunkers. Near Landing Zone (LZ) Sierra on infamous Mutter's Ridge, the new "Mike" Company used very little indirect fire while neutralizing four separate defense complexes in a row. This all had happened in I Corps' most challenging arena—the area just west of "Leatherneck Square." The following award citation explains how the absence of external firepower had helped Mike's commander to get the job done with minimal casualties.

On 13 March 1969, First Lieutenant [Edwin C.] Kelley [Mike's new company commander] was directed to retake Landing Zone Sierra . . . which had been previously abandoned by friendly forces and was subsequently occupied by a North Vietnamese Army force entrenched in well-fortified bunker complexes. After personally leading a reconnaissance patrol to within 100 meters of the hostile emplacements without detection, . . . [he] formulated his plan of attack and initiated an aggressive assault on the enemy positions. During the ensuing protracted engagement, . . . [he] directed his company in the destruction of a series of four bunker complexes *without the aid of air support and*

Figure 8.2: No Amount of Shelling Will Quiet Every Defender MG
(Source: http://search.usa.gov public-domain image from this url: http://www.history.army.mil/images/artphoto/pripos/amsoldier/5/1945_Brazilian.jpg)

with only limited artillery fire. When monsoon weather precluded helicopter resupply, he instructed his Marines in the employment of captured North Vietnamese Army weapons and grenades for a final assault against the remaining hostile fortification, thereby enabling his company to seize the objective and establish defensive positions. During the night, the Marines were subjected to a series of probing assaults, which increased in intensity until the early morning hours when the enemy penetrated a sector of the [Marines'] perimeter. . . . He fearlessly led a bold counterat-

tack resulting in the defeat of the North Vietnamese Army force. [Italics added].[9]
— Navy Cross citation for E.C. Kelley

When This Is Possible, What Else?

If nine infantry squads require no artillery or air support to overpower a well-planned defense matrix, then a single squad doesn't need any to punch its way through a hasty encirclement. That means no firepower umbrella would be necessary to safely blanket a region with squad-sized American additions to host-country police outposts. Should that outpost come under attack by a vastly more numerous force, the U.S. component has only to break through any cordon and E&E (with its indigenous counterpart in tow) over to the nearest neighbor.

9 Monitoring a Large Area with Few U.S. Forces

● How can a huge area be watched by a small contingent?

● In which ways can GIs become better force multipliers?

Dispersed forces must be well integrated into population.

(Source: http://search.usa.gov public-domain image from this url: http://www.jointbasemdl.af.mil/shared/media/ggallery/hires/AFG-100730-021.jpg)

Sgt. Price's Message to the Establishment

Sgt. Toney Price is the only legendary Marine with whom this author has become personally acquainted. Mentioned briefly in Chapter 7, his exploits will again be featured in the next chapter. As with many of the WWII heroes, Sgt. Price needed periodically to "interpret" his orders to both hurt the foe and safeguard his troops. Though slightly wounded while constantly in search of contact, he steadfastly refused a second Purple Heart or any heroism award. The best testament to his contribution is what his men thought despite his highly authoritarian leadership style. Ron Baldwin, Brad

Bennett, and Kirk Hauser all visited their former NCO's deathbed several times before traveling long distances to attend his graveside memorial service. Though having never served with Price, his neighbor Steve Molumby became so attached that he lent him a house and then became his longtime (and unpaid) caregiver. Toney ended up with no family or money, just a few former subordinates and loyal friends. Though suffering terribly from Chronic Obstructive Pulmonary Disease (COPD), he never complained. Among his favorite sayings was, "Marines don't whine."[1]

The lesson behind Toney's adventures is the leeway that all squad leaders must have if their parent unit is to stick with a 2GW agenda in a 4GW environment. Locally "interrogating" prisoners

Figure 9.1: All Captives Sent Off to Higher Headquarters
(Source: "Vietnam Suspects," by Ronald A. Wilson, U.S. Army Center of Military History, posted October 2006, from this url: http://www.history.army.mil/art/A&I/1006-4.jpg)

isn't one of them. (See Figure 9.1.) Nor is overcontrolling squad members. But, retreating when the ammunition runs out definitely qualifies. Another is enough spirit to put their squads into a position where withdrawal may become necessary.

America can't win WWIII by every squad leader now emulating some past legend. Toney's way worked in Vietnam because his adversary wasn't expecting it. Though his tactical techniques are certainly worthy of any modern-day squad leader's toolbox, his leadership style would fail miserably for most. A full-blooded Choctaw and well acquainted with the woods, Toney was uniquely qualified for the job of Combined Action Company (CAC) squad leader. Then, under extremely difficult conditions (to include too little headquarters support and a military that had ill-prepared him to face good light infantry), he did what he had to. When that military finally does evolve tactically at the small-unit level, the average squad leader will be able to do just as well. He will have riflemen so skilled and self-assured as not to need much supervision; and most of his combat control will come through prior squad practice of common-scenario solutions.

The Price Ambush

Toney Price had come from a Marine Corps that was still enamored with the early WWII idea that all squad members must function as one in battle. That is most easily accomplished by all rehearsing standard maneuvers and then doing exactly as told by their squad leader when the enemy shows up. As such, Price was a strict believer in "immediate-action drills" to counter any enemy ambush. Though good at building teamwork and quickly counterattacking, these drills allowed for little member deviation—for an unexpected obstacle (like a fallen tree)—from the assigned-quadrant maneuver. Because almost identical procedures became common practice in the post-Vietnam Marine Corps, Toney's will not be described here. What will be described is his favorite way of applying 2GW tactics to a 4GW scenario in which his squad was always outnumbered and outgunned. That involved an ambush of his own. (See Figures 9.2 and 9.3.)

From that ambush's extremely tight formation, one could conclude that Price liked to overcontrol his personnel. However, as CAC compound leader, he probably had one Marine fire team and two or three Popular Forces (PF) fire teams in the average patrol. And not all the accompanying Leathernecks were from an infantry background. Though Price spoke some Vietnamese, the language barrier alone would have required closer-than-normal supervision.

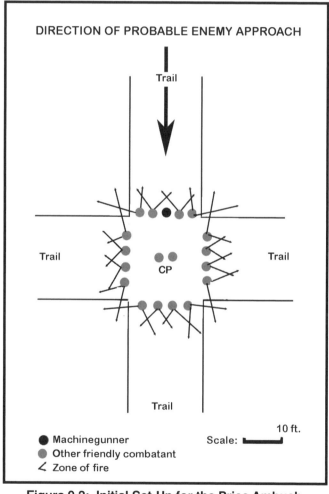

DIRECTION OF PROBABLE ENEMY APPROACH

Trail

Trail

Trail

CP

Trail

Machinegunner

Other friendly combatant

∠ Zone of fire

Scale:

10 ft.

Figure 9.2: Initial Set-Up for the Price Ambush

Because Price was 2GW trained, he would have associated victory with killing enemy personnel. VC fighters, political commissars, and tax collectors all qualified. As only enemy maneuver elements (and no civilians) traveled after dark, everything on the nighttime trails would have been fair game. So, Price reasoned the best way

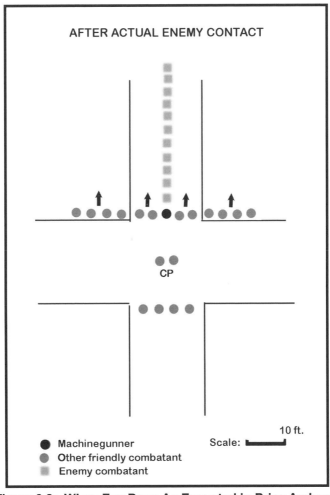

Figure 9.3: When Foe Does As Expected in Price Ambush

to kill a bunch of VC was to put his tiny element squarely in their path. He did so by massing at the very center of an active trail junction. Having previously encountered opposition at the rear of a linear ambush, he opted for a tiny perimeter instead. It would more easily handle trouble from an unexpected direction. With

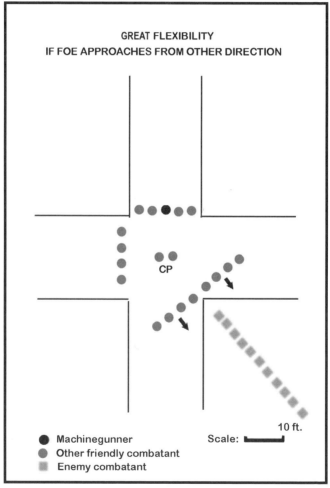

Figure 9.4: When Foe Does Something Else during Price Ambush

four fire teams, a perfect "square" provided the best fields of fire. (Look again at Figure 9.2.) Anything round would have delayed the counterstroke. Because all fire teams were virtually within arm's reach of his Command Post (CP), Toney could quickly move them around to meet any threat.[2] (See Figure 9.4.) Then, those fire teams attacked.

Sgt. Price was enough of a student of guerrilla warfare (and mismatched encounters in general) to know that the name of the game was "hit and run." Constantly in search of a firefight, he would occasionally have to go back to his compound for more ammunition. It was this unfortunate reality that was eventually to get him into trouble with his 2GW-oriented headquarters. For them, moving backwards in the face of the enemy was never acceptable. That went double for squads and below. Price's willingness to be labeled a coward to keep his people from being overrun is what may have endeared them to such a strict disciplinarian.[3]

The tactical genius in Price's ambush formation was twofold: (1) its potential for actually damaging the foe; and (2) its ability to move in any direction (including backwards). Price knew that troop confidence and morale largely hinged on actually hurting somebody. So, he gave his men the following instructions. Anyone could trigger the ambush as long as the intruder was "about to touch the front-sight blade of that person's rifle." The fields of fire would not just overlap horizontally, but also vertically. Some men were to aim six inches above the ground, while others would shoot at "bellybutton level." As long as their target remained visible, all were to empty as many 20-round clips of ammunition as possible. Each had an M-16 capable of "fully automatic" fire. On ambush patrol, Price encouraged everyone to carry extra magazines. Canteens or anything else that might make noise were not allowed. Toney's reason was not only more stealth, but more firepower. Of course, such a lavish expenditure of rounds made it more likely they would eventually run out. Then, contact would have to be temporarily broken, so that the unit could go back to the compound for resupply. While winning the Medal of Honor (MoH) on Iwo, Cpl. Stein had made no fewer than eight trips back to the beach to get more ammunition for his customized machinegun.[4]

To cover any withdrawal, Price had a preregistered mortar tube back at his compound. Thus, all ambushes were totally based on firepower, with the only means of resupply included.[5] That's the inconsistency with present-day U.S. operating procedure that must be addressed if a large area is ever again to be controlled by tiny well-dispersed elements. 2GW depends almost entirely on firepower, but makes no provision for keeping it going under such circumstances. A nearly suicidal bayonet attack like that of Chamberlain at Little Round Top or the 7th Cavalry at Ia Drang (in "We Were Soldiers"[5]) cannot be the only solution. Neither should playing it safe to keep from incurring the dilemma. For Sgt. Price's badly outnumbered patrols, every enemy contact was an ambush of sorts that might require their own pullback. That's close to the infantry version of UW.

In the minds of Price's personnel, the possibility of rearward movement was more than adequately offset by his highly aggressive agenda. Should one of his patrols pass through an untriggered enemy ambush, he and his people would circle around and "wipe out" its occupants.[6] Toney knew the best way to keep his men safe was to stay on the offensive. Another former enlisted Marine shared this view. His name was Puller.

Unfortunately, the political cost of taking any casualties whatsoever has, over the years, largely obscured this reality. As a result, most of the fighting is now done during the day by large U.S. units and from overly protective formations. Not surprisingly, the tiniest of their subordinate elements then have trouble developing much interest in close contact.

Thus, the Price example makes a good start to this chapter. As a CAC squad leader, he was responsible for monitoring a large area with few U.S. personnel. To do so, he had first to keep his Marines, PFs, and tiny compound from being annihilated. But, that was yesterday, and this is now. Today's problem is how scores of more average squad leaders could—in a heavily populated area and with less force—equal Price's accomplishments from a 4GW standpoint.

Still Missing from the Pentagon Equation

To oversee a large overseas area, a small expeditionary force would have widely to disperse tiny "self-sufficient" elements. Remotely watching that area from above makes too little provision for the coincidences of everyday life. Such a coincidence could easily result in collateral damage.

Due to the increased danger of operating alone, each element must be able to hold out until dark (without any outside assistance) and then either hide in place or exfiltrate an encirclement. Remotely induced escape lanes take a broader supporting-arms umbrella than is generally available. Often, they will only postpone the danger. Where several elements get into trouble at once, there aren't enough air assets to go around. Instead of parent-unit firepower, each element needs only a few secretive maneuvers.

Asian armies have no trouble blanketing a region with token forces. The Communists have made a veritable art form of it. Often, not all participants remain fully identifiable as troops.

If All U.S. Troops Must Be Uniformed

If all deployed GIs must remain fully visible to the indigenous population, then their reason for being in a foreign country must be something other than military occupation. The People's Republic of China (PRC) has long used the cover of "U.N. peacekeeper" or corporate security guard.[7] The West need not be as devious. Its troops could possibly be of help to local authorities.

If the Pentagon remains unwilling to grant Geneva Convention rights to non-uniformed rebels, then it should try the law enforcement approach to controlling them. Aren't insurgents normally in violation of some local law? Why shouldn't they subsequently be treated as criminals—and then afforded all the considerations of international jurisprudence?

A New Mission

As most large wars now grow out of 4GW circumstances, the best way to prevent one is through societal assistance. That assistance can run the gamut from food aid to security. It is the latter at the local level that offers the best opportunity for military involvement. Without enough local security, there can be no functioning national democracy.

Often, those manning police outposts need a little paramilitary training just to keep from being overrun. With a squad of U.S. infantry instructors in residence, those outpost personnel could

85

Figure 9.5: Marines Practice Law Enforcement on Shore Patrol
(Source: FM 19-95B1/2 [1978], cover)

better do their job. Among other things, that job would be to limit the amount of Islamist, Communist, and organized-crime activity in their zones of responsibility.

Because light-infantry expertise is tantamount to proficiency in the subcategories of supposed basics, junior GIs could most easily achieve it. To do so, they must be given a reason to subdivide and then experiment with what they think they know. Within Asia, instructing militiamen has provided this reason. During the Vietnam War, no NVA soldier was allowed to move south before up to a year of educating northern militiamen.[8] When they finally did go south, many were tasked with teaching VC.

As with any reputable police department, outside instructors would have immediately to inform their own headquarters of any excesses in suspect treatment. The operative word here is "suspect." Until, each "person of interest" has been formally tried in a court of law, he or she must be afforded all the protection of international police protocol.

The Use of Deadly Force

All American police officers have sworn to protect not only innocent bystanders, but also the criminals themselves. They are not allowed to preemptively fire at any suspect who does not pose a lethal threat to themselves or another person. Assumed innocent, that suspect has constitutional rights. Only, in those rare instances where the person is known to be a convicted murderer, can he (or she) be fired upon while running away.[9] Nor, upon capture, can any suspect be tortured or otherwise coerced/tricked into providing a confession.

Truly professional policemen do not maltreat detainees. They know that any information obtained through forceful interrogation is likely to be a fabrication. That every suspect is innocent until proven guilty guides their every action. For a supposed evildoer to be punished, a jury must first be convinced (with enough hard evidence) that he is guilty. That takes more investigative ability than the average American rifle squad can muster. Still, its members learn easily.

No Paradigm Shift Necessary

The members of the above-mentioned American squad must only consider themselves on overseas garrison duty in a somewhat dangerous location. U.S. Marines used to draw "Shore Patrol" or Military Police (MP) duty at every foreign port or station. Whether they or their Army brethren still do is unimportant. Just making sure all standing orders are being obeyed during a "fire watch" or "sentry post" constitutes law enforcement experience. (See Figure 9.5.)

Therefore, to function effectively in this new "expeditionary" role, the members of each American squad need only a garrison mind-set. Instead of "closing with and destroying an enemy," they must now "serve the public," "enforce local discipline," and "protect the weak." No longer is every stranger a potential target for their weapon. When annihilation looms, they must only become hard to find.

First, the squad must develop enough self-sufficiency to operate semi-autonomously. For the U.S. Marine Corps (USMC), that is already encouraged by its new doctrine.

10 New Doctrine Calls for More Squad Autonomy

- What year did USMC switch to its new tactical doctrine?
- What is that doctrine called?

Germans have relied on squad strongpoints since WWI.

(Source: "The Red Bull in the Winter Line," by D. Neary, poss. ©, from url: www.nationalguard.mil/resources/photo_gallery/heritage/hires/Red_Bull_in_the_Winter_Line.jpg)

This Situation Up Front Is Fluid

For U.S troops in contact with the enemy, the competition between headquarters control and frontline necessity can be a problem. Time and detail will always be lost between any forward occurrence and the commander's notification. An additional delay occurs between his decision and the lead element's ability to execute. In the meantime (with an opportunistic opponent), frontline conditions have changed. Then, lowest-echelon initiative may be necessary to save the day. Such initiative can come close to violating orders. Here's a story from Vietnam to illustrate the point.

Only Reason "CAC 10" Wasn't Annihilated

There's an old Marine Corps saying: "Following every order can be as dangerous as not following any." The movie "Windtalkers" explores this paradox in depth. Its star—a dedicated Sergeant—first loses every man in his squad while obeying an order to defend a "worthless piece of swamp." Then, he amends "his duty" by extracting under heavy fire (instead of shooting) a Navajo code talker about to be captured.[1] There are no hard and fast rules on which combat orders strictly to obey, but most are fairly general in content. This lack of detail lends itself to a little "interpretation." The average combat commander would gladly settle for the "spirit" of most orders being applied to frontline reality, but subordinates can't tell which in advance. Sgt. Toney Price guessed wrong. The following incident resulted in court martial charges that were later dropped. It also earned him the undying respect of his subordinates.

Sgt. Price did not hold position when his patrol was attacked by company size or larger VC unit while on patrol outside of CAC 10 (village of Gio La), half way between Phu Bai and Hue City in fall of 1966. . . .

. . . I was scared shitless of "Sarge" Price . . . [when] sent out to CAC 10 to operate the only 81mm mortar tube in the compound. Sarge Price made it clear . . . that if I thought I was coming out to the Ville to get a "f . . . ing tan" as he put it, I was badly mistaken. I would have to take my turn on night guard duty as well as go on patrol like everyone else. In fact, one of the first nights I was on perimeter watch it was like zero dark thirty, and I could feel my eye balls getting heavy and about to slam shut when I felt this presence behind me. I turned to find "Sarge" Price looking down on me, and all he said was "Bennett you fall asleep in my hole, and I will kill you myself."

On this particular day, Sarge Price took me aside and told me he was taking a patrol over to a village out in the rice paddy on the other side of Highway (1) that intelligence said had VC moving in and out of it. We went over map lay outs and plotted fire missions that he might need along the way. I told him that if he needed fire support to call me and I would call fire command in Phu Bai and get permission to fire. At which time, he said, "Bennett if my troops are under fire and I call you for support, you had better have those rounds on target or . . . not be here when he [I] got back off patrol." I said [I] understood.

As the patrol advanced on the village I followed by radio. At one point Sarge Price said he was close to the ville and could see many VC moving and taking up positions in a trenchline out in front of the ville. He radioed me the location and moved his patrol to the left toward a rice patty dike when all hell broke loose. They opened up with heavy and light machine guns as well as AK-47s and RPGs. I could hear the firefight raging as Price called in positions to fire on. At one point, he told me I was on target and to fire at will traversing left and right until he told me to stop. The contact went on for 40 to 50 minutes, . . . [and] I fired almost every HE [high explosive] round I had. He then called and told me he was all but out of ammunition but had been told by headquarters in Phu Bai to hold [a] blocking position until they could get another unit out of Phu Bai to sweep the ville. I told him I didn't know how much more cover I could give him since I was almost out of rounds. It was at that time he decided to withdraw his patrol back to our compound and that I would provide covering fire with everything I had left.

When his patrol came back through the wire, he walked up to me and put his fist in my chest and said, "Hell of a job Bennett." That was the only time I ever remember him giving me a compliment. The very next day he was called in to headquarters in Phu Bai for not holding his position, and we didn't see him again out at CAC 10.[2]

— Brad Bennett, 81mm mortar man at CAC 10
18 February 2013 e-mail

Another member of CAC 10 describes the same incident in slightly more graphic terms. When out there alone, American squads need more than just orders from some rear-area location. Sergeant Price was almost punished for what should have earned him a medal.

Unlike guerillas *[sic]* or members of highly decentralized Asian armies, who have no direct communication with higher command, our troops are burdened with decisions made from higher ups in air conditioned shelters miles away from the noise and horror of combat.

. . . I was in 81 mortars [with 1/4] during Operation Oregon. We were pinned down by .51 cal. machineguns and the requests for 81 mortar fire missions was denied, 'friendlies in the area'. The decision was made by an officer not even on site. We were the friendlies. Azimuths, coordinates, aiming stakes were not needed, for we could see the enemy.

I was also attached to a CAC . . . unit, where we lived with the locals. Our unit consisted of one squad of Marines and a platoon of local militia. Sgt Price . . . had what he called 'Immediate action drills'. Every Marine knew exactly what to do for any likely situation, whether it be getting ambushed on patrol or just encountering an unknown civilian. Safe areas were assigned for reassembly if needed. We knew the complete area around us—houses, yards, pig sties, alley ways—and could tell from the reactions of the locals if something was amiss. We became a family, not only among ourselves, but within the village. When the squad was on patrol, fire missions were always denied because of 'friendlies in the area'. On one occasion, I heard small arms fire and knew the patrol was being hit. When they called for support, mortars were quickly raining down for them, because I did not request permission to fire. The patrol was given orders to hold position (a decision made by someone not there), but luckily the radio transmission was garbled, for they only had 7 bullets among them. A Sgt Price tactical technique was used instead—after contact, never stay in area, enemy knows where you are. Go to designated safe area and set up 360 [degree] ambush.

There's a different—more productive—way for the squad leaders to create their own control. It's by pre-rehearsing their men in the solutions to likely situations. Sometimes called "immediate-action drills," or "tactical techniques," these football-like plays then ensure enough cooperation to outmaneuver the foe. The result will be more timely decisions made on the spot using common sense.[3]

— PFC Ron Baldwin, member of CAC 10

That Uniquely American Trait

Young U.S. males are so free spirited that any predisposition toward independent activity must be suppressed in boot camp. Instead of restoring this loss of initiative, commanders generally settle for more team orientation. That, after all, makes their enlistees easier to lead. (See Figure 10.1.)

Yet, every young American rifleman's initiative is still missing and could have been usefully "resurrected" in a more-mission-friendly format. Because Chinese society is more group oriented, its military recruits get initiative training. That's why the Carlson

Figure 10.1: Teamwork Has Been Stressed for U.S. Recruits
(Source: "Return from Pre-Invasion Maneuvers," by Olin Dows, U.S. Army Center of Military History, from this url: http://www.history.army.mil/art/dows/61_27_45.JPG)

Raider way of building self-confidence could help U.S. boot camp graduates to reacquire the most productive parts of their former initiative. Whether low-ranking combatants have initiative or not, the problem with short-range combat remains the same. The "orders" they are asked to execute will seldom mesh with the latest changes to their situation. When fighting a stratified Western opponent, this makes little difference. But, when up against a "bottom-up" and "frontline-oriented" Eastern foe, it makes all the difference in the world. America's most recent deployments should have reaffirmed this truth. Without allowing more personal initiative from the lowest echelons, no American ground force may ever again be able to handle an Eastern foe. How many more examples of this do U.S. traditionalists need? To date, U.S. forces have failed to win a clear-cut victory in Korea, Vietnam, Iraq, and Afghanistan—not to mention several interim skirmishes. Westernized French, Russian, and—for a while—even Mainland Chinese forces have experienced similar problems in those same regions. The thread throughout is the greater degree to which Asian Communists and militant Muslims have regularly afforded semi-autonomy to the tiniest of their contingents.

Marines' New Doctrine Promotes Squad Opportunism

The U.S. Marine Corps officially changed its doctrine from Attrition to Maneuver Warfare (MW) in 1986. MW specifically calls for more squad autonomy.[4] Only then can control be sufficiently dispersed to fully practice "soft-spot tactics." Unfortunately, that part of the transformation never took hold. Part of the problem was in its predisposition toward minimal force.

> Maneuver warfare can be thought of as military *judo*. It is a way of fighting smart, of out-thinking an opponent you may not be able to overpower with brute strength.[5]
> — William S. Lind, father of MW in America

The Pentagon has routinely managed fire superiority—at least at the point of attack—against all comers. Why shouldn't every subordinate element do likewise? Another problem has been confusion over what role, if any, squad formulas still play in MW. Its doctrinal predecessor did almost everything "by the numbers" at that level. That's one of the reasons it was so plodding and predictable.

> Tactics combines. . . techniques and education. Techniques are things you can do by formula, like . . . unit battle drills. Excellence in techniques is very important to maneuver warfare. But good techniques are not enough. . . . Tactics includes the art of selecting from among your techniques . . . for [unique circumstances]. Education is the basis for doing that.[6]
>
> — William S. Lind, father of MW in America

Though hard to determine from the MW literature, the concept of "technique" mostly applies to squads and below. Like a football play or "stunt," a technique takes only a little practice and then muscle memory to execute. Maneuvers involving more than 14 people generally take a leader's continual oversight. There are few exceptions. Some whistle-announced company maneuvers were attempted by the People's Liberation Army (PLA) at the Chosin Reservoir, but with limited success.[7] Because techniques are vital to MW, so are squads.

> The rifle squad currently occupies a relatively minor place in Marine Corps tactical thought. . . . Considered in terms of maneuver warfare, this attitude is disastrous. Because [it] is often the point of contact, squad must be able to react to changing situations and to seek out enemy weaknesses. This initiative, for which the German *Stosstruppen* became famous, demands that the squad assume a primary tactical role.[8]
>
> — William S. Lind, father of MW in America

Perhaps the biggest reason the Marines have been unable to embrace this essential part of their new doctrine is "trying too hard to comply with headquarters directives." What the Commander in Chief has just ordered must certainly be followed, but his directives seldom involve enough detail to restrict lower-echelon initiative. Only when subsequent levels in the long chain of command restrict the options of those below, do problems arise. Though ostensibly loyal, their strict "attention to orders" then stifles the ingenuity that could have led to a frontline victory. Most heavily affected are the people at the very bottom of the hierarchy—those in contact with the enemy.

> Maneuver warfare involves radical decentralization of control.... Speed requires that decisions be made at the lowest possible level.[9]
>
> — *Maneuver Warfare: An Anthology*

Figure 10.2: The Real Problems Began after Dark
(Source: "The Rock of the Marne," DA Poster 21-42, U.S. Army Ctr. for Mil. Hist., from this url: url: http://www.history.army.mil/images/artphoto/pripos/usaia/Rock.jpg)

To see how vital "deep" delegation of authority can be, one has only to study the 1917 German Army. To salvage WWI, it attempted defensive formations and offensive maneuvers that were both non-linear. The latter can best be described as long-range "infiltration tactics"—something made possible by ultra-savy assault squads. Those squads had at their disposal a MW technique so powerful as secretly to penetrate any Doughboy line, however well established. (See Figure 10.2.)

[S]mall units organized for independent action—the famous storm troops—used favorable terrain, particularly culverts, ravines, and similar features, to bypass enemy strongpoints and penetrate into the depth of the enemy positions.[10]
— *Maneuver Warfare: An Anthology*

The WWI Germans had discovered that MW was largely about what highly proficient squads do. By the end of 1917, they had NCOs in charge of all ground attack spearheads,[11] and elastic-defense forts.[12] Their heritage of deeply delegating authority then made WWII a much more tenuous proposition for the Allies.

Only by decentralizing control over training and operations will U.S. Marines ever fully reap the benefits of their new warfare style. After a few well-tested headquarters' guidelines,[13] each company must be allowed to run its own squad experiments.

Winning in combat requires . . . excellence in techniques.[14]
— FMFM 1-3, foreword

Asian Communists Can Already Do MW at Squad Level

By the late 1930's, Mao Tse-Tung had supplemented the "lone-squad maneuvers" of WWI Germany with others just as helpful.[15] The WWII Marine Raiders then found his way of training combatants so insightful as to copy it. Something similar may still prove useful to U.S. forces. (See Figure 10.3.)

With Deeper Delegation of Authority Comes Responsibility

If U.S. squads are to become more autonomous, they must have

Figure 10.3: Troops More Often Alone in Lower-Intensity 4GW
(Source: http://search.usa.gov public-domain image from this url: http://www.cslib.org/connector/CONNectorPics/pr10000409_646e5c29b.jpg)

carefully conceived ways of regulating their own behavior. What they would need adequately to fulfill this responsibility comes next.

Part Four

Such Squads Take Raider-Like Training

"HONOR IS EVERYTHING. . . . YOU FOLLOW THE RULES OF WAR FOR YOU—
NOT FOR YOUR ENEMY. YOU FIGHT BY THE RULES TO KEEP YOUR
[OWN] HUMANITY." — WWI AERIAL-TACTICS INSTRUCTOR

(Source: As told to Franz Stigler by commanding officer Gustav Roedel during pilot training, from "Individual Honor," Jacksonville Daily News (NC), 3 February 2013, p. 27)

11 Riflemen Need More Than — Rules of Engagement

● How many moral issues can frontline fighters consider?

● Might additional rules squelch their tactical opportunism?

Collateral damage happens during more than just airstrikes.

(Source: http://search.usa.gov public-domain image from this url: http://arts.gov/sites/default/files/styles/large-620/public/02-seeing-3.jpg?itok=FbwYM16z)

Each Rifleman's Conscience Can Be a Valuable Asset

Atrocities happen in war. Many, like Vietnam's My Lai incident, are most easily understood as group evil. Enemy excesses are recalled, a leader's "pep talk" gets misinterpreted, someone shoots indiscriminately, and others follow suit. Not every U.S. teenager has arrived at the same understanding of right from wrong by the time they enlist. In the rush to replace his self-centeredness with a little teamwork, that ever-so-logical check of his conscience seldom occurs. Even with the semi-autonomous Raiders, Carlson's "Ethical Indoctrination" had nothing to do with moral-mistake prevention. It was only to ensure that each man knew why he was fighting.[1] It was a way of "giving conviction through persuasion."[2] Out of this indoctrination, Carlson had hoped to give his men more confidence in themselves, their leaders, and their peers. It primarily focused on why each unit member's efforts counted, so that no one would think his job more important than another. Out of this mutual respect and confidence would supposedly come the ability "to work together wholeheartedly, without fear or favor or envy or contempt."[3]

Unfortunately, captives pose somewhat of a problem for fast-moving guerrillas, so Carlson's attempt to seize "the moral high ground" had ignored a significant crest. He had made no attempt to invoke each Raider's conscience. If that indoctrination had been more specific about how to treat enemy soldiers and civilian sympathizers, Carlson's experiment might have succeeded. Would the same limitation apply to any 2GW strategy in which success is measured through numbers of enemy killed?

When That Interior Voice Must Be Encouraged

Orders are to help GIs work toward some mutual goal, not to control their every thought. For most, intentions are good, and all temptations accompanied by an internal warning. The following "letter to the editor" of a civilian newspaper serving a military community shows what can be accomplished when low-ranking people are allowed to follow their inner calling.

"[T]his is the true story of a man who had a voice in his head and what happened because he followed that voice. You can be the judge of whether or not it was destiny.

Raymond M. Clausen Jr. was his name, but if you met him he'd tell you to just call him Mike. Mike's tour in Vietnam had ended and he was back home, but he put in a request for another tour in Vietnam. His mother asked him why he wanted to go—why? Mike said, "I have something I got to do. I don't know what, but there is something I still have to do."[4]

— Letter to the Editor,
U.S. Daily, 3 December 2013

Mike Clausen didn't know what he was supposed to do, only where he was supposed to do it. So, he returned to Southeast Asia for a second dangerous tour. While displaying the below-mentioned initiative, he was still just a Private First Class (PFC) in the USMC.

On Jan. 31, 1970, Mike and that little voice met head on. This is where the little voice, fate, destiny and free will converged. Participating in a helicopter rescue mission to extract elements of a platoon that had inadvertently entered a minefield while attacking enemy positions. Mike skillfully helped the pilot to a landing in an area cleared by one of several mine explosions.[5]
— Letter to the Editor,
U.S. Daily, 3 December 2013

The letter writer then gives every reader the chance to arrive for themselves at the source of Mike's impending heroism. Destiny (a close facsimile of fate) has already been mentioned. So, now he discusses "free will," because Mike did not have to obey that little voice in his head. He would have intentionally to decide whether to do something so scary as to be virtually suicidal.

Mike's decision was to jump out of the helicopter and run into the minefield under heavy fire and help carry out the wounded. Other Marines would follow in his footsteps out of the minefield to the waiting helicopter. He did this six times until all Marines were safely aboard and the helicopter lifted off. Mike saved 18 Marines that day.[6]
— Letter to the Editor,
U.S. Daily, 3 December 2013

Then, the letter sender helps his audience to see how "providence" (namely, the will of God) may have had something to do with Mike's unselfish act. That would certainly explain his non-PFC-like degree of involvement.

For the actions Mike took that day, he would receive the Medal of Honor.

And still the question remains: Was it destiny, fate or providence? . . .

Mike Clausen passed away in 2004. His actions on that one day would define Mike for the rest of his life as a hero. Nearly every Marine who ever met Mike, liked Mike—he was the living example of "no man left behind."[7]
— Letter to the Editor,
U.S. Daily, 3 December 2013

Mike Clausen's admirer is not the only one to share this conclusion. A U.S. Navy Chaplain creed attributes all heroism to the Holy Spirit within.[8] Even for non-churchgoers, there is proof. James Corbett, India's legendary man-eating-tiger hunter of the 1930's and 40's, had only two near fatal encounters during some 250 victories. This was no small feat, as his equally wily foe could see after dark and liked to sneak up on a pursuer from the back. Luckily, Corbett is one of the very few "one-on-one" *aficionados* to have fully chronicled his methods. He once shot—by slowing reversing the direction of his rifle—a tiger that was about to pounce on him from the rear.[9] When asked about his apparent "sixth sense," Corbett gave his Guardian Angel the credit.[10] In other words, his last-second insights had come from a providential messenger. All of the world's major religions agree that every human being (to include each U.S. rifleman) has a Guardian Angel.[11]

A Few Rules of Engagement Can't Insure Ethical Behavior

The average American-expeditionary-force Rule of Engagement goes something like this: "Don't damage any historic or religious structure." It can also establish the commensurate response for a standoff attack. During the Vietnam War, enemy sniper fire was not to be countered with U.S. artillery rounds. In Iraq, there were other guidelines for how close an unchecked civilian vehicle could get to an Allied fort or convoy. Mostly unit oriented, Rules of Engagement almost never address the moral pitfalls of a lone soldier's encounter.

To avoid moral error, U.S. squad members will need more structure than that. A good place to start is what to do while attacking when a defender starts to raise his hands. As all U.S. assault troops are expected to keep up with their formation, wounded defenders also pose a problem.

Figure 11.1: Minuteman "Swarm" Forced British Back to Boston
(Source: "Shot Heard 'Round the World'," by D. D'Andrea, poss. ©, from url: http://www.nationalguard.mil/resources/photo_gallery/heritage/hires/Concord_Bridge_full.jpg)

The Popular Misconception

Within a hierarchical Western society, citizens easily conclude that any real progress takes centralized control. For Colonial militiamen, that wasn't true in 1775. (See Figure 11.1.) For American grunts, it wasn't true in WWII. (See Figure 11.2). Nor is it true for today's Pentagon community. In fact, by simply working together, small groups of ordinary GIs can do amazing things without any leader at all. Perpetuation-bent bureaucracies don't like to admit this. For many of their higher-ranking members, the whole idea of loosely controlled squads performing more proficiently (and ethically) than those that have been closely supervised is preposterous. Still, it might be possible, if those squad members were to acquire a greater degree of self-discipline.

Figure 11.2: Omaha Landing Saved by Individual Determination
(Source: "First Wave at Omaha," by Ken Riley, poss. ©, from these two urls: http://www.history.army.mil/images/reference/normandy/pics/blue-Gray.jpg and
http://www.nationalguard.mil/resources/photo_gallery/heritage/hires/First_Wave_at_Omaha.jpg)

No Major Changes Really Necessary

Perceptions are everything when it comes to modifying a long-standing practice. Most U.S. military "lifers" see new recruits as virtually defenseless through too little exposure to organizational procedure. Only after enlistees get the "hang of things" will they become survivable. On the other hand, most American citizens believe their sons and daughters about to receive the best tactical instruction in the world. Neither of these perceptions is totally accurate.

Young Samuel Jones is not nearly as helpless as his careerist "superiors" would like everyone to believe. After the constant threat of close combat in high school, he is already predisposed toward light-infantry knowledge. Though his post-recruitment training will be topnotch, it will mostly focus on how to operate the next generation of equipment.

Yet, both segments of society can agree on one thing. The current crop of American teenagers could use a little more self-discipline. Luckily, most who join the infantry see the same thing as a way to further their goals.

Personal Honor Still Important to Most Riflemen

The Pentagon (formerly War Department) has seldom admitted to losing any enemy engagement, however small. As a result, the average U.S. military officer has a somewhat inflated opinion of his organizational capabilities. For he who designs the maneuver and then moves to the center of a rather formidable formation, the idea of personal honor differs substantially from that of his teenage bodyguards. (See Figure 11.3.) They can't afford as much "positive

Figure 11.3: Those Who Want "Death before Dishonor"
(Source: "Preventive Maintenance: M203 and M79 Grenade Launcher" [pamphlet reproduced by U.S. Army Armor School, Fort Knox, KY])

thinking." Should an immediate adversary suddenly appear with an edge in locale or ability, they could end up making the ultimate sacrifice. As such, they have fewer dreams of fame or fortune. In its place is a rather fatalistic sense of responsibility toward nation, family, and friends. These young infantrymen must match the exploits of grandfather and father, or die trying. Most dream of performing some heroic deed on the "field of honor."

Sadly, chance encounters are more often decided by relative location and skill than determination. Just as honorable as dying in a blaze of glory would be a long-term commitment to handling hellish conditions with as much compassion as possible. That way, every unit more easily maintains its assigned end strength. Such a thing would take more mental conditioning.

Advantage of Operating within Each Man's Capabilities

In life, one generally does better by sticking to what he already knows. Many a novice plumber or turkey chef has come up a little short by charging off into uncharted territory. Among the world's biggest armies, the Asian Communists best compensate for this axiom of human behavior. Their troops are not trained as, or expected to be, killing machines—only the often secretive facilitators of strategic damage. Such a mission fits more comfortably with their Creator's design. A well-endowed Western adversary relies heavily upon materiel, so why not destroy it?

To achieve this more difficult goal, the Asian commander works harder at discovering his people's capabilities than "training" them. If possible, he fights only when they enjoy some edge in experience or terrain over their adversary. After a well-attended diversionary attack (or threat thereof), he risks only a few sappers to destroy the Western unit's supply dump, indirect-fire capability, and/or command element. For him, most terrain holds little inherent value. When continuing to fight offers no strategic import, he withdraws.

Meanwhile, Western armies generally march to the sound of the guns. After chasing a barely achievable standard of physicality, killer instinct, and marksmanship, their troops are supposedly ready for anything. The nobility of their mission is then to bring out the best of their instincts. Unfortunately, that's not what always happens. If too badly hurt in the initial stages of an assault, these same Western troops will sometimes resort to less than commendable behavior with any enemy combatants they happen to encounter.

Both ways of fighting are valid, but one has a better chance of establishing individual squad momentum across a wide frontage. It will also instill more deep-seated self-confidence within each squad member.

Most Americans have trouble accepting the above paradox. They have been exposed to too much bureaucratic "spin" over the years. Atheistic Communists do have less regard for human life than practicing Christians, but not less for each junior billet assignment. Because of the Oriental "bottom-up" way of doing things, most Red commanders place more value on what each individual soldier can contribute to the whole. Only that soldier will notice a minor crack or deception in the opponent's frontline array. If too little appreciated, he will keep that crucial intelligence to himself. To be fully opportunistic in battle, U.S. military commanders must pay more attention to what each lowest-ranking infantryman is experiencing.

Detailed Plans of Little Use When Balloon Goes Up

Military men have long known that the best laid plans go out the window as soon as the first shot is fired. That's because the foe is not cooperating. Getting off the beach at Omaha took more than orders and procedures. (Look back at Figure 11.2.)

That most desperate of days was saved by lowest-echelon initiative. Junior officers may have decided which part of the German formation to penetrate, but it was enlisted survivors of many units who then worked together to make it happen. Most could have easily avoided the responsibility. But those brave lads had suddenly become mission-essential, and they reveled in this chance to finally make a difference. Each of today's enlistees is similarly capable of ensuring that their unit will always enjoy battlefield success. In fact, he will stake his life on it. (Refer again to Figure 11.3.)

How Asian Troops Are Prepared

As the Asian commander learns what his troops can do from

Figure 11.4: GIs Learned What Red Troops Can Do in Korea
(Source: "Corporal Hiroshi N. Miyamura, Korean War," by George Akimoto, U.S. Army Ctr. for Mil. Hist., from this url: http://www.history.army.mil/art/A&I/0507-3.jpg)

peacetime battledrills, he permits modification to those drills. Among other things, each nonrate will be shown how to fight like a guerrilla. That gives him enough confidence to work alone when participating in squad-level mobile warfare (much the same as the "maneuver" variety). During the Korean War, Americans were first exposed to this light-infantry threat. (See Figure 11.4.)

How Oriental Armies Instill Riflemen with Self-Discipline

Because East Asians (most notably the Chinese) culturally avoid any outward sign of emotion, they may appear to be insensitive or stupid. Compounding this illusion is a long tradition of battlefield deception. The reigning grand master of *ninpo* calls his advanced form of *ninjutsu* the working knowledge of one's own subconscious level of thinking.[12] Thus, at the heart of the East Asians' preparation for individual combat is most likely the "self-discipline" that comes from a pair of *ninpo* precepts.

Protecting one's own body, mind, and spirit [13]
Developing a benevolent heart *(jihi no kokoro)* [14]

The first would be quite foreign to any Western grunt. He has been taught to subordinate all personal concerns to unit mission. All parts of his body must do as instructed despite any danger to its aggregate. His mind is not to complain. And the only "spirit" with which he has been linked is unit morale. However, strictly following orders is quite different from surviving to fight another day. Though American commanders mostly focus on the here and now, they wouldn't mind having a rifleman who could keep coming back for more combat.

Staying alive is largely a personal endeavor. It involves obeying instincts, carefully operating, taking cover, staying low, and the occasional pullback. Those kinds of things don't mesh well with the average American scheme of attack. That scheme has been based on standardized maneuvers in which every rifleman remains upright and moving forward. The smaller the maneuver element, the more survival anomalies are possible, but any crawling during an assault is definitely out.

Most thought-provoking within the first *ninpo* precept is the reference to "spirit." While its full meaning is well beyond the purview of this study, there are still hints. The NVA way of building momentum through gradually larger victories would suggest a link to each individual's self-confidence.[15] Thus, the precept may not be referring to the spirit that comes from doing one's part to facilitate the whole, but rather the kind that grows out of overcoming personal adversity.

A Benevolent Heart?

Asians are mostly non-Christian. As such, they should have difficulty developing a "benevolent heart." Christian Westerners would more easily see how such a thing might help the frontline fighter. Yet, neither is the case. To the average U.S. citizen, warfare and morality pose competing visions. For an interface, he must go to the Navy Chaplains' creed. Everything good can be associated with the Holy Spirit, so a benevolent heart would be as well. Such a heart would then help its recipient to display more audacity in combat. That men fight more for each other than any other reason has now been widely acknowledged. How well each unit does in war may thus depend on the mutual regard between junior enlisted personnel. This would explain why badly outnumbered Marines were still able to overcome the most advanced defense to date on ideal terrain at Peleliu, Iwo Jima, and Okinawa.

Atrocity Prevention through Individual Responsibility

Though most Americans are highly ethical, their elected leaders will sometimes settle for political expediency. Yet, both worry that poorly supervised GIs might do something unseemly in battle. The best hedge against it is peer pressure. That's what Evans Carlson was able to harness in his Marine Raiders through the Maoist traditions of freely admitting mistakes and then "working together" to correct them.[16]

Service personnel who honestly believe themselves to be "the best in the world" at all things would have no reason to improve. If they had somehow incurred a light-infantry deficiency, not knowing "how to use small forces" would then cause their commanders to lose some battles (according to Sun Tzu).[17] That no part of the U.S. defense apparatus has ever felt it necessary to closely study an opponent makes the odds even worse.

When you are ignorant of the enemy but know yourself, your chances of winning and losing are equal.[18]
— Sun Tzu

When combined, these two "errors of pride" would then forfeit the tactical advantage in over half of the encounters. But, with all that firepower, who cares?

Mao's 8th Route Army soldiers and later those from the NVA weren't that handicapped from a maneuver standpoint. They were encouraged to constructively criticize superiors and check battle chronicles for accuracy. By forcing unit commanders to face up to their tactical mistakes, they helped to preclude a recurrence. Though far from perfect, Asians see great shame in professing to have won a battle they really didn't. That they are continually improving lowest-echelon performance is what makes their units so difficult to defeat at short range.

If U.S. riflemen were instilled with more self-discipline, apprised of all ethical issues, and then allowed to operate in semi-autonomous groups, they could also police their own behavior. That's how best to utilize an isolated squad, because any decision-making delay could lead to its annihilation. Of course, its members would require somewhat different preparation. (See Figure 11.5.)

Figure 11.5: Welcoming a Chance at His Full Warrior Potential
(Source: "Supertroop," by SFC Darrold Peters, U.S. Army Center for Military History, from this url: http://www.history.army.mil/art/Peters/Supertroop.jpg)

12 — Creating a More Proactive Fighter

● Why must U.S. riflemen become more assertive?

● How will this help their unit?

How can so heavily laden a soldier properly defend himself?

(Source: http://search.usa.gov public-domain image from this url: http://www.history.army.mil/art/varisano/1.44.91.jpg)

When It Helps to Be a Self-Starter

A wartime GI won't always have buddies nearby. Anywhere within Asia, an enemy soldier can suddenly pop up out of the ground at arm's reach. Then, that GI will be totally on his own. Whether he survives will depend on his own degree of agility and initiative.

How One Becomes More Proactive As a Fighter

Upon rejoining the Corps in early 1942 (after three years of "broken time"), Clyde Thomason became one of the first members of 2nd Raider Battalion. Then, Lt.Col. Evans Carlson picked him to lead the advance element on the Makin Island Raid. It was at the Butaritari garrison on that raid that Sgt. Thomason earned the MoH.[1] His exceedingly bold approach to offensive operations is what set him apart from his peers. Where there was any doubt as to what to do next, he attacked (and generally at the head of his tiny contingent).

This willingness to take action where there were no specific instructions to do so had not come from his first tour (1934 to 1939) with the Marine Detachment of the Flagship for the Asiatic Fleet.[2] There, strict attention to orders and procedure would have been the name of the game. Sgt. Thomason's opportunistic spirit had come from his Maoist Raider training. Within 2nd Battalion, all hands had been encouraged to combine more of a "say-so" on things with additional self-discipline. That's because of what Mao had learned about leading guerrillas. Widely dispersed elements can't be as closely supervised.

> With guerrillas, a discipline of compulsion is ineffective.[3]
> — Mao Tse-Tung quote
> 1941 *Marine Corps Gazette*

Through that nontraditional Raider instruction, the former ceremonial guard had become preeminently qualified to lead the advance element of the first U.S. ground attack in the Pacific after Pearl Harbor.

For conspicuous gallantry and intrepidity at the risk of his life above and beyond the call of duty while a member of the Second Marine Raider Battalion in action against the Japanese-held island of Makin on August 17–18, 1942. Landing the advance element of the assault echelon, Sergeant Thomason disposed his men with keen judgment and discrimination and by his exemplary leadership and great personal valor, exhorted them to like fearless efforts. On one occasion, he dauntlessly walked up to a house which concealed an enemy Japanese sniper, forced in the door and shot the man before he could resist. Later in the action, while leading an assault on enemy position, he gallantly gave up his life in the service of his country. His courage and loyal devotion to duty in the

face of grave peril were in keeping with the finest traditions of the United States Naval Service.[4]
— MoH Citation for Clyde Thomason

All told during WWII, former Marine Raiders and their Navy Corpsmen were to be awarded seven Medals of Honor, 136 Navy Crosses, 21 Distinguished Service Crosses, 330 Silver Stars, 18 Legions of Merit, six Navy and Marine Corps medals, three Soldier Medals, 223 Bronze Stars, and 37 Letters of Commendation.[5] The most junior of the MoH recipients was PFC Henry Gurke of 3rd Battalion. While serving alongside Carlson's Raiders at a Bougainville trail junction, he had absorbed the full blast of a grenade so that a foxhole partner could continue to repel a major enemy assault with his Browning Automatic Rifle (BAR).[6]

More Self-Discipline Takes Nontraditional Preparation

Times have changed. Now, most U.S. recruit training attempts only to expose everyone to enough organizational procedure to meet an acceptable norm for personal behavior. That's not the same as self-discipline.

Self-discipline is what riflemen need when temporarily devoid of instructions or supervision. It's that on which their very survival may depend when no small-unit leader is around to give them advice. There are decisions that a "snuffy" must make in combat that are not covered by his general orders or Rules of Engagement. Among them are whether or not to shoot, investigate a ground anomaly, or report something of strategic import. Someone too used to following orders may lack the self-confidence to make such a decision. That's why the authoritarian boot camp routine must at some point be augmented by something a little more open ended. Most logical would be stations at which every fledgling fighter could freely react to various "signs" of enemy activity. Among them might be buried ordnance, human footprints, or charging assailants.

The hard truth on this subject is undeniable. Fully to contribute to an infantry or special operations unit, the average U.S. boot camp graduate will need a more assertive attitude. For advice on how to instill such an attitude, one only has two choices—rely on foreign military experience or resort to untested philosophy. The army of a bottom-up culture would be the best place to look. Here's how many Oriental commandos have learned to control their own minds (as confirmed by the *ninpo* capabilities on Table 2.1.)

Maintaining a "can-do" attitude
    Gaining inner strength through meditation [7]
    Transcending fear or pain [8]
Enhancing one's awareness of surrounding conditions [9]
    Concentrating on sensory impressions [10]
    Paying attention to detail [11]
    Reading the thoughts of others [12]
    Perceiving danger [13]
    Making decisions [14]
    Visualizing the task to be accomplished [15]

Of course, a freewheeling Yank would require limits to that much introspection. Though a more assertive spirit might pay him and his unit great dividends (like more battlefield longevity), that spirit must be carefully focused to maintain the same level of teamwork.

A More Proactive Attitude through Inner Awareness?

Before venturing into the rather risky realm of "inner awareness," every American infantry trainee will have be told to limit any free thinking to matters of a tactical or ethical nature. Then, whether or not he agrees with his subsequent orders, he will have to either follow them or defend why he didn't at a possible court martial. All other regulations and procedures will have to be happily embraced.

The junior infantryman's frontline location gives him unique insight into any "bottom-up" (criminal or Asian) foe's tactical intentions. He will also be the first to see how much U.S. collateral damage has actually been done. Carlson had afforded his men *"Gung Ho* Sessions" on maneuver after an "Ethical Indoctrination." That's why everything tactical and ethical is now deemed worthy of lowest-echelon input.

There are ways other than the *ninpo* suggestions to attain the proper mix of proactive behavior. The Oriental concept of "meditation" might be replaced by each rifleman's religious-prayer preference. Pain can be transcended through heavier physical workouts, and fear through always doing the hardest thing first. Any other "inner strength" recipes should be carefully weighed by a board of experienced trainers. Yet, few would argue with the value of more closely monitoring ongoing circumstances.

Becoming More Alert to One's External Environment

To better handle personal challenges, one needs less focus on the boss's agenda and more on other conditions. The *ninpo* list perfectly applies. It advises additional attention to four things: (1) one's own senses; (2) situational detail; (3) the adversary's body language; and (4) possible threats. Among the best ways to encourage them are lonely missions: (1) point man; (2) mantracker; (3) stalker; (4) short-range infiltrator; or (5) listening-post occupant. When there is no time to rehearse, every task will go much smoother after a mental walk-through of its steps (the Asian concept of "visualization"). More "aggressor duty" during unit training will also increase one's situational awareness.

Overall Confidence Building

Being constantly told what to do, treated like a new guy, or given unrealistic tests of physical coordination do little to increase a young grunt's confidence. More reliance on his own abilities would come from single-handedly dealing with multiple assailants. That may be why Asian Communist guerrilla training so heavily focuses on being outnumbered. American trainers may want to mirror this approach.

For the U.S. rifleman quickly to shoot every assailant would be only part of the equation. Marksmanship alone will not save him from every scenario. If suddenly surprised when alone, he may have to maneuver through rocks, ditches, and bushes just to avoid enough enemy fire to rejoin his friends. A snuffy who had fallen asleep during a rest stop in Vietnam later came huffing and puffing into his company's sweep formation. He then declared that VC had been chasing him around the base of the hill for the last 30 minutes.[16] Thus, the kind of "surety" that U.S. riflemen experience is mostly of

119

Figure 12.1: U.S. Army Assault in Korea
(Source: "The Borinqueneers," by Dominic D'Andrea, poss. ©, from these two urls: http://www.defense.gov/specials/heroes/images/pedro_art.jpg and http://www.nationalguard.mil/resources/photo_gallery/heritage/hires/The_Borinqueneers.jpg)

the organizational or unit variety. They have mistakenly assumed that their parent-unit's firepower advantage will always keep them safe from harm.

Of course, a certain tactical certitude also arises out of being able to secretly approach aggressor lines during training. However, a soon-to-be-outnumbered assault squad requires more of its members than that. Enough resolve to do what comes next in actual combat would take everyone's personal experience with a rapid succession of assailants. Sneaking up on many times one's number and then actually closing with them are quite different evolutions. That's why Carlson encouraged so much "one-on-one" grappling. By allowing each man to "freelance" the first feint, he instilled both initiative and confidence.

Instead of being part of the impervious "steamroller" that most Americans imagine, all U.S. infantry platoons have been regularly exposed to devastating fire while assaulting enemy strongpoints. (See Figure 12.1.) Once the riflemen gained entry, neither was their consolidation as orderly as the manuals purport. The inside-the-perimeter fight can be as chaotic as at Bunker Hill in 1775. (See Figure 12.2.) Then, the rote ways of shooting or stabbing all who resist may not be enough. What if there are just too many of them? Each American would then be better off emulating a "Brave Heart" warrior—to handle a rapid succession of foes while still watching his buddies' backs.[17] The first requires a well-practiced portfolio of personalized feints. The second takes thinking while otherwise occupied—something that enough "muscle memory" can help.

Figure 12.2: A Few Standardized Moves Not Good Enough Now
(Source: "Bunker Hill: 17 June 1775," Army Ctr. of Mil. Hist., posted Aug. 2002, from this url: http://www.history.army.mil/images/artphoto/artchives/2000/avop06-00_2.jpg)

Figure 12.3: "Heavy" Infantrymen Seldom Get to Operate Alone
(Source: "Knock Knock," by SFC Darrold Peters, from U.S. Army "War on Terror Images," at this url: http://www.history.army.mil/books/wot_artwork/images/8a.jpg)

Expanding the Basics

A more self-sufficient U.S. rifleman would need additional training in several categories: (1) microterrain appreciation; (2) harnessing the senses; (3) night fighting; (4) guarded communication; (5) discreet force at close range; (6) combat deception; (7) escape and evasion; and (8) one-on-one encounters.[18] These subsets of the existing "shoot," "move," and "communicate" basics have been fully described in *The Tiger's Way*. They have come from a list of *ninjutsu* capabilities (Table 2.2 of that reference). The current study will only add what *ninpo* and Carlson's Raiders have to say about the seventh and eighth.

Escape and Evasion

Chapter 2's *ninpo* chart mentions the need for sometimes hiding. Its sole entry is "making oneself invisible." Most vital to that end is mind-set. A man with no particular agenda will more easily look like the bush he is in. Next most important is going where not expected—like up a tree, beneath a pile of farm feces, or into air-conditioning duct work.

That same chart mentions four ways of eluding capture: (1) establishing a false exit; (2) leaving a trail and then retracing one's steps; (3) doubling back after being bypassed; and (4) abruptly terminating one's trail. The last can be sometimes accomplished by moving along barely abutting tree limbs, rooftops, and telephone lines.

One-on-One Encounters

Carlson's men had learned to watch for an onrushing attacker from any direction, and then—through some ruse—to gain the upper hand. His command guidance had been "always to expect the unexpected."[19] At Camp Pendleton's Raider School, freewheeling aggressor duty took the place of "canned-exercise" participation.[20] Such would have meant being constantly outnumbered. Yet, nowhere in the Carlson literature is there any mention of a Raider jousting with several opponents at once or even in quick succession. This, after all, is the Asian way of doing things. Many *karate "katas"* have been specifically designed for that purpose. Thus, each prospective U.S. rifleman could usefully practice brawling with more than one opponent. (See Figure 12.3.)

A More Specific Recommendation

There is little difference between an Asian martial-arts move and U.S. self-defense procedure. Though bayonet thrusts have been the GIs' normal bill of fare, a few *jujitsu* throws, knife slashes, and *karate* strikes have occasionally been thrown in. Yet seldom, if ever, has each rifleman been asked to combine moves while countering more than one adversary.

123

During pugil stick training, that GI was allowed blows of his own design against a single peer. But, that very productive medium will have to be expanded to include more than one foe, non-rifle contacts, and occasional feints. The new "strategic rifleman" must be able to handle several opponents in quick succession. That's because he and his light-infantry squad will be regularly penetrating large enemy positions where a counterattack is likely.

As in *karate,* an established sequence of moves can prove awkward for some people. That's why each rifleman must be allowed to choreograph his own combinations. To achieve the most power and speed, his kicks, hits, and ducks must all flow naturally from the ones before. After a right-leg kick, for example, he may have trouble launching a good left hook.

Why So Pressing a Need for This Kind of Training?

Since the redesignated Raider units of WWII, all U.S. infantrymen have been of the "heavy" or "line" variety. They are "unit oriented" and almost totally dependent (for their own protection) on issued weaponry. Such troops get far too little exposure to the finer points of close combat. Periodically attempting a few is not enough. Strictly to obey doctrine, they need a three to one edge even to close with an adversary. This gives most an inflated opinion of their own personal capabilities. Such group-and-firepower-oriented GIs are at considerable risk anywhere within Asia (to include its Southwestern regions). If they run out of ammunition, are too slow to reload, or have malfunctioning rifles, they may have trouble handling not only a *ninjutsu*-trained counterpart, but anyone more used to operating alone.

The American people and their duly elected leaders have too easily fallen for the arms manufacturers' claims. Properly "taking care of the troops" involves more than issuing them the latest equipment. That equipment can be heavy, fickle, and difficult to master. If required for every mission, it can become a veritable millstone around each participant's neck.

If Asian Armies Train This Way

Since 1947, all Asian Communist riflemen have been expected

to take on many times their number under all conditions.[21] Possibly because of every unit's guerrilla warfare requirement, this enlisted tradition has given Red army commanders more of an opportunity to practice Mobile (or MW) at the small-unit level.

Whether the Chinese, North Korean, and Vietnamese armies require each rifleman to practice close combat with several opponents in a row is not known for sure. That most martial-arts disciplines do, makes it likely. Asian enlistees almost certainly receive more one-on-one self-defense preparation than the average GI. Just prior to WWII, Japanese units were holding pre-dawn bayonet fights between individuals. Besides a wide assortment of unarmed hits and tosses, Asians also know how to dodge an incoming blow. Within the *ninjutsu repertoire* alone, rolling, leaping, and tumbling away from an opponent all have received enough attention to rate their own Japanese names.[22]

The Most Logical Modification to the U.S. Curriculum

American riflemen will be carrying small arms in the assault, so they must learn to handle—with bayonet or rifle butt—more than one defender. Such skills can be most easily acquired incrementally—through gradually more opposition or circumstantial difficulty.

Initially, only the "mock-rifle" contact should be expanded. All surprise-oriented assault technique depends upon no shooting noise, so what's the harm in training with a piece of wood. After pugil sticks have been issued, each American would try to counter two assailants in close succession, then two at once. All the while, he would be improvising his own moves. Some time later (like the pre-WWII Japanese had done), each GI would—with actual rifle, sheathed bayonet, and padded gloves—then fight a similarly armed pal. Their first joust would be during the day, and their second at night. Both would wear helmets with face guards and refrain from full contact. Any blows to the face or neck would earn the transgressor a trip through a peer flogging line.

Then would come the unarmed portion of the self-defense class—that involving *jujitsu*. After practicing some of the basic moves, each rifleman would counter one onrushing defender, then two in quick succession, then three.

Whether each American gladiator could actually defeat more

Figure 12.4: A More Self-Contained Rifleman
(Source: "Cav Trooper," by SFC Darrold Peters, U.S. Army Center of Military History, from this url: http://www.history.army.mil/art/Peters/Cav%20Trooper.jpg)

than one adversary would not matter. Simply that he had tried would give him more confidence. After such a modification to U.S. training, the old "bulldog jaw" (so common in WWII photographs) will make a comeback. It should serve its new owners well during their more isolated assignments of the 21st-Century. (See Figure 12.4.)

13 What Such Riflemen Add to Unit Power

- What can a squad do after rehearsal and reconnaissance?
- How can it reconnoiter while already attacking?

The unit member ready for any challenge.

(Source: Courtesy of Cassell PLC, from "Uniforms of Elite Forces," © 1982 by Blandford Press Ltd., Plate ?, No. 8)

What More Proactive Personnel Do for Their Parent Unit

Among the greatest of infantry-unit challenges is attacking a well-fortified enemy position. Ideally, this is done by surprise. But, sometimes that surprise gets compromised during the final assault. Then, it may be too risky to pull back and try again later from some other direction. This is when opportunistic and aggressive small-unit leaders come in handy. Two former Raiders were to demonstrate these qualities while winning the MoH in some of the most vicious fighting of WWII. The first was William Gary Walsh on Iwo Jima; the second was Richard Earl Bush on Okinawa.

Walsh had been with Carlson's Raiders (probably as a junior NCO) on Guadalcanal and Bougainville.[1] His award citation identifies the high point he was attacking on Iwo Jima only as Hill 362. Unfortunately, that island had three pieces of terrain with this same elevation. If 362A had been Walsh's target, it was part of Defense Line Number Two, honeycombed with underground passageways,[2] and attached by tunnel to Nishi Ridge.[3] (See Maps 13.1 and 13.2.) Nishi was the high ground that ran along Iwo's northwest coast. It sported no fewer than 100 camouflaged entrances to caves that were all linked below ground. One passageway ran for 1,000 feet,[4] so it must have been the one back to 362A. But, if Walsh's hill had been 362B or 362C, it was farther to the northeast and part of Defense Line Number Three. The former was covered with a "dense network of interconnected caves and pillboxes."[5] (Look closely at Map 13.3.) Either way, Walsh had helped his parent unit to capture one of the most strategically important bastions on the island.

For conspicuous gallantry and intrepidity at the risk of his life above and beyond the call of duty as Leader of an Assault Platoon, serving with Company G, Third Battalion, Twenty-seventh Marines, Fifth Marine Division, in action against enemy Japanese forces at Iwo Jima, Volcano Islands, on 27 February 1945. With the advance of his company toward Hill 362 disrupted by vicious machine-gun fire from a forward position which guarded the approaches to this key enemy stronghold, Gunnery Sergeant Walsh fearlessly charged at the head of his platoon against the Japanese entrenched on the ridge above him, utterly oblivious to the unrelenting fury of hostile automatic weapons and hand grenades employed with fanatic desperation to smash his daring assault. Thrown back by the enemy's savage resistance, he once again led his men in a seemingly impossible attack up the steep, rocky slope, boldly defiant of the annihilating streams of bullets which saturated the area, and despite his own casualty losses and the overwhelming advantage held by the Japanese in superior numbers and dominate position, gained the ridge's top only to be subjected to an intense barrage of hand grenades thrown by the remaining Japanese staging a suicidal last stand on the reverse slope. When one of the grenades fell in the midst of his surviving men, huddled together in a small trench, Gunnery Sergeant Walsh in a final valiant act of complete self-sacrifice, instantly threw himself upon the deadly bomb, absorbing with his own body the full and terrific force of the explosion.

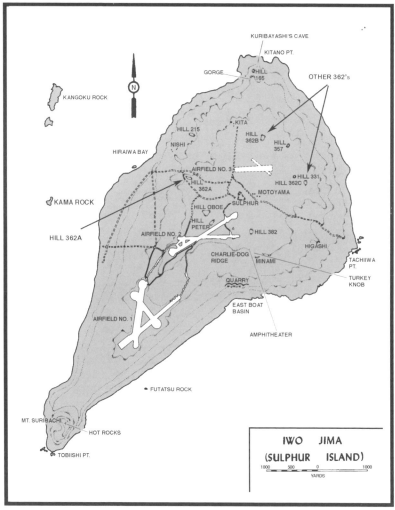

Map 13.1: The Terrain of Iwo Jima
(Source: "Closing In: Marines in the Seizure of Iwo Jima," Hist. & Museums Div., HQMC, 1994, p. 5)

129

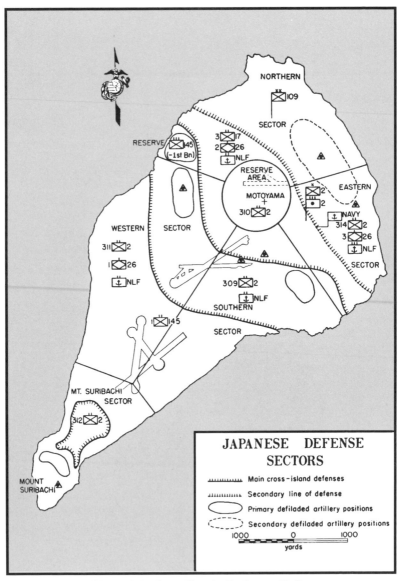

Map 13.2: Iwo Jima's Defense Belts

(Source: "Closing In: Marines in the Seizure of Iwo Jima," Hist. & Museums Div., HQMC, 1994, p. 8)

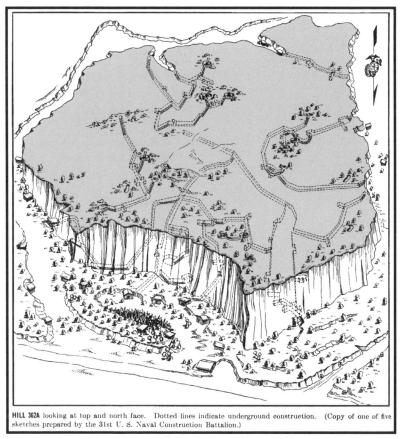

HILL 362A looking at top and north face. Dotted lines indicate underground construction. (Copy of one of five sketches prepared by the 31st U. S. Naval Construction Battalion.)

Map 13.3: Extent of Tunneling beneath Hill 362A
(Source: "Iwo Jima: Amphibious Epic," by Lt.Col. Whitman S. Bartley, from sketches by 31st Naval Construction Battalion, Hist. Branch, HQMC, 1954, p. 140)

Through his extraordinary initiative and inspiring valor in the face of almost certain death, he saved his comrades from injury and possible loss of life and enabled his company to seize and hold this vital enemy position. He gallantly gave his life for his country.[6]

— MoH Citation for William Gary Walsh

131

The other MoH winner—Richard Bush—was a Corporal with 1/4 when the award was won on Okinawa, but probably of lesser rank in 1st Raider Battalion.[7] More traditional in format than the two Maoist Raider battalions, 1st had still adopted many of Carlson's methods while under the command of Samuel Griffith.[8] Fully to see what Bush did at Mt. Yae Take will take a short synopsis of the battle. (See Map 13.4.)

> [T]he 4th Marines moved out at 0830 with 3/29 on the left, 2/4 on the right and 1/4 in regimental reserve. An artillery, air and naval gunfire bombardment preceded the attack. The advance met light resistance and the Marines reached their initial objective, a 700 ft. high ridge before noon. During this time, 1/4 moved to the right rear of 2/4 to guard the open right flank. Later, Company C and then Company A were committed to assault a ridge 1,000 yards to the right front of 2/4.
>
> Following preparatory fire, 2/4 and 3/29 resumed the attack to seize a ridge 1,000 yards to their front. As the advance continued, resistance steadily increased, ambushes from cleverly concealed positions were frequent and many officers became casualties. The Battalion Commander of 1/4 was the victim of such an ambush in an area that had been quiet for hours
>
> Company G of 2/4 came under heavy rifle, machine gun, mortar, and artillery fire about 2,000; and in a few minutes Company E was hit with the same fires. Naval gunfire was called down on an artillery position that had been spotted and the gun fell silent. Ignoring the heavy casualties suffered by Company G, 2/4 pressed the assault and took the ridge utilizing an envelopment from the right. By 1630 2/4 and 3/29 were on the regimental objective, and 1/4 occupied the high ground to the right. Close contact was then established between all units.[9]
>
> — Dan Marsh's Raider Page, USMCRaiders.com

It was during 1/4's assault against this right-hand ridge that Bush did almost precisely what Walsh had done on Iwo—repeatedly rally his men up a steep slope against massed enemy fire. By then absorbing grenade blasts, both demonstrated just how committed they were to group and mission. Of course, many non-Raiders had

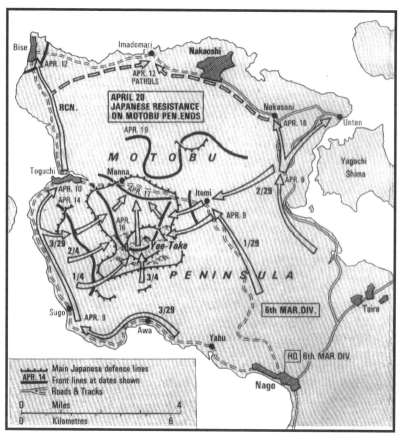

Map 13.4: 4th Marines' Assault at North End of Okinawa
(Source: "The Final Campaign: Marines in the Victory on Okinawa," by Col. Joseph H. Alexander, Marines in WWII Commemorative Series, HQMC, 1996, p. 17)

also continued to attack at other places in the Pacific. Still, few had done so under more trying conditions or for as important a target. All Raiders had been continually apprised of overall strategy, so these two must have figured out how crucial their assignments were to taking those islands. That's why both were to risk so much. Together, their citations should be adequate proof of what Raider-trained individuals can add to unit power.

For conspicuous gallantry and intrepidity at the risk of his life above and beyond the call of duty as Squad Leader serving with the First Battalion, Fourth Marines, Sixth Marine Division, in action against Japanese forces during the final assault against Mt. Yaetake on Okinawa, Ryukyu Islands, April 16, 1945. Rallying his men forward with indomitable determination, Corporal Bush boldly defied the slashing fury of concentrated Japanese artillery fire pouring down from the gun-studded mountain fortress to lead his squad up the face of the rocky precipice, sweep over the ridge, and drive the defending troops from their deeply entrenched position. With his unit, the first to break through to the inner defense of Mt. Yaetake, he fought relentlessly in the forefront of the action until seriously wounded and evacuated with others under protecting rocks. Although prostrate under medical treatment when a Japanese hand grenade landed in the midst of the group, Corporal Bush, alert and courageous in extremity as in battle, unhesitatingly pulled the deadly missile to himself and absorbed the shattering violence of the exploding charge in his own body, thereby saving his fellow Marines from severe injury or death despite the certain peril to his own life. By his valiant leadership and aggressive tactics in the face of savage opposition, Corporal Bush contributed materially to the success of the sustained drive toward the conquest of this fiercely defended outpost of the Japanese Empire and his constant concern for the welfare of his men, his resolute spirit of self-sacrifice and his unwavering devotion to duty throughout the bitter conflict enhance and sustain the highest traditions of the United States Naval Service.[10]
— MoH Citation for Richard Earl Bush

GI Longevity Now More Important Than Winning?

Nobody likes losing GIs in wartime, but every victory has its price. Should that price never even be risked, then all casualties will be in vain.

In America's most recent conflicts, casualty avoidance seems to have been the overriding concern. Any losses, however slight, became political liabilities. Was the Pentagon's edge in technology supposed to keep all ground troops from getting hurt? Wherever opposition forces mix with local citizenry, this won't often be possible.

There are two ways to limit U.S. losses: (1) fight defensively until having to abandon the war effort; or (2) learn to fight more effectively. It has often been said that the best defense is a good offense.

How Other Nations Have Handled This Challenge

Since WWII, America's traditional adversaries (most notably the Asian Communists) have been developing ways to negate the West's advantage in electronic surveillance and precision firepower. One is widely to disperse all forces—with the infantry squad as base element. Some contingents may remain visible, but their overall number becomes too great and individual size too small for much of a Western response. Though never admitted by the Pentagon, such a strategy is possible in the Orient because of the greater skill and initiative of squad members.

The Germans are not the only European army to have also discovered how tiny, isolated groups can survive. So had the Finns in their war against the Russians.

> The Finns [in their 1939 repulse of a Russian invasion] emphasized the ability of the individual soldier as the basis for the effectiveness of the whole unit. . . .
>
> The theory was that overcoming even a strong enemy could be done with localized attacks and flanking movements; thus splitting the enemy to digestible pieces and destroying them in detail, one pocket at a time. Precisely how this was done was left for the [small-unit] commanders on the ground to decide.[11]
>
> — "Finland in the Winter War," by Ville Savin

So What's the U.S. Military's Problem?

Until semi-autonomous infantry squads can be conclusively proven to suffer acceptable losses in conventional combat, such a dispersion strategy will never be tried by the Pentagon. That's a real shame, because nothing larger than a squad can secretly close with a prepared enemy position. If defenders are never to be surprised, then how can an American commander hope to maintain momentum?

A partially isolated U.S. squad would need more assertive GIs. Though its current members might be good as a group at various forms of maneuver, they won't always be working at arm's length. Just for that squad safely to move around an active battlefield would take the further deployment of march security teams. The proactive rifleman's greater initiative makes him more capable of two-man security and infiltration missions. Having already come to blows with a string of assailants in training, he more easily survives a close encounter. Only still missing from his *repertoire* are the finer points of working alone—like how to escape an encirclement without any outside support. Over the years, U.S. troops have rarely been allowed to hide, much less move backwards. Their Communist counterparts haven't been so unfortunate.

Proactive Fighter Development

With Raider-like leadership, U.S. riflemen will automatically become more assertive. Effectively to apply such assertiveness, they will need more of an advanced skill set. Besides mastering the additional "basics" of Chapter 12, each must come to appreciate what a single rifleman can accomplish on his own. Table 13.1 gives examples and their source. Before the Posterity Press initiative, this degree of tactical detail had either been edited out of the mainstream manuals or so seldom mentioned as never to reach the average grunt. Contrary to popular opinion, the infantry field is the most complicated of all occupational specialties. As such, true proficiency lies in its details.

What That Proactive Fighter Makes Possible

Every unit is still the sum of its parts. Better parts create more opportunity for group advancement. That's how more skilled and assertive riflemen will finally make semi-autonomous U.S. squads possible.

America's German, Japanese, Chinese, North Korean, and North

CONVENTIONAL APPROACH EXPERTISE

LAND NAVIGATION BY TERRAIN ASSOCIATION	LHYCHAPT9
WALKING POINT	LHYCHAPT10,TWCHAPT15
STALKING	TWCHAPT17
RURAL MANTRACKING	TWCHAPT16
URBAN MANTRACKING	TTCHAPT14

CONVENTIONAL ATTACK EXPERTISE

FIRE AND MOVEMENT SKILLS	LHYCHAPT12
SHORT-RANGE INFILTRATION	LHYCHAPT20, DDCHAPT15(PP230-235)
NOT GETTING SHOT IN THE CITY	LHYCHAPT24(PP318-319), MTCHAPT15(PP288-294)
SAFELY MOVING INTO A CONTESTED BUILDING	HSCHAPT9(PP166-184)

CONVENTIONAL DEFENSE EXPERTISE

PROPERLY POSITIONING A FIGHTING HOLE	LHYCHAPT21(PP262-267)
CLEARING FIELDS OF FIRE	GWCHAPT18(PP236-240)
STANDING WATCH FROM A FIGHTING HOLE	DDFIGS15.9THRU15.12
MANNING A NIGHTTIME LISTENING POST	TWCHAPT13
FULLY DEFENDING A ROOM	LHYCHAPT23(PP308-311)

UNCONVENTIONAL WARFARE KNOWLEDGE

FIGHTING LIKE A GUERRILLA	GHCHAPT3(PP29-35), TJCHAPTS14 &16
ESCAPE AND EVASION IN ENEMY COUNTRY	DDCHAPTS18 & 20, DDCHAPT19(PP311-325), DDCHAPT21(PP350-362)
BAREHANDED TANK KILLING	TWCHAPT13(PP73-75)
DECEPTIVE MEASURES ON DEFENSE	GWCHAPT18(PP235-238), EECHAPT15
OFFENSIVE RUSES	GWCHAPT17(PP228-233)

CODE:
LHY, THE LAST HUNDRED YARDS, ISBN ISBN 0963869523
TW, THE TIGER'S WAY, ISBN 0963869566
MT, MILITANT TRICKS, ISBN 0963869582
TT, TERRORIST TRAIL, ISBN 0963869590
DD, DRAGON DAYS, ISBN 096386954X
TJ, TEQUILA JUNCTION, ISBN 0963869515
HS, HOMELAND SIEGE, ISBN 9780981865911
EE, EXPEDITIONARY EAGLES, ISBN 9780981865928
GW, GLOBAL WARRIOR, ISBN 9780981865935
GH, GUNG HO, ISBN 9780981865942

Table 13.1: Proactive Rifleman Prerequisites

Vietnamese foes have all used lone squads to spearhead attacks and anchor defenses. They did so for a reason. That reason was not any shortage of localized manpower or firepower. They maximized their level of surprise and flexibility that way. All had discovered more momentum possible through strings of small-unit attacks than large-unit offensives. Low-risk scenarios are easier to find for small units. If any squads got into trouble, they had only to pull back. A large unit can't as easily disengage.

Stormtrooper squad penetrations of successive Doughboy lines (those arrayed in depth) in 1918 involved a series of quick yet still deliberate attacks. An attack is said to be "deliberate" when it has been reconnoitered and rehearsed. While those WWI Germans had certainly practiced their famous assault technique, they were then forced to rely on "recon pull" for any information about subsequent targets. The forward-most elements of spearhead squads provided that service. The *ninpo* ways of building confidence would also help with impromptu reconnaissance: (1) sensory impressions; (2) attention to detail; (3) reading sentry thoughts; and (4) perceiving danger.

Though the infantry squad as a whole may practice a maneuver, each member must then separately rehearse his composite roles or later resort to visualization. Through *ninpo*, visualization becomes a viable option.

How a single pair of proactive riflemen might penetrate a prepared enemy position will be discussed shortly. But first must come a more aggressive way for the average U.S. squad to patrol. Without any "strategic riflemen" onboard, it would have to stay well clear of any fortified area.

How More Assertively to Run a Patrol

For whatever reason, the U.S. military has become far too conservative between major offensives. Then, even its security patrols can have trouble making contact in areas that are literally awash with enemy personnel. This lack of "commensurate response" makes American morale as hard to maintain as momentum.

There is a way to run more aggressive squad-sized patrols without ever having to ignore a sighting or expose a flank. The patrol's size is what makes this possible. A dozen or so soldiers can easily transition between column and line formations. Where something out of the ordinary is spotted to one side of the march route, those dozen most powerfully approach it on line. That only takes all hands simultaneously turning to the right or left and moving forward. Then, another facing movement away from traveling single file in the original direction, that new line formation is also one of the strongest defensively. (See Figure 13.1.) With an automatic-weapons man on either end and other occupants alternately pointing inboard and outboard, it can easily withstand a human-wave assault from any direction.

Only the patrol's flankers would be, in any way, challenged by all main-body members moving to the side. The farthest flank security team could have no set way of rejoining the line formation. It would have to move—by the most direct route—to a place alongside the rest of its fire team. (See Figure 13.2.)

Figure 13.1: Quickest Attainable Defensive Formation
(Source: "Army/Marine Clipart," U.S. Air University, from this url: www.au.af.mil/au/awc/awcgate/cliparmy.htm)

After investigating the anomaly, all squad members have only to pivot in place to resume their original column formation. Sideways offsets of 100-200 yards would scarcely bother the average "land navigator" of a preplanned route.

Of late within U.S. forces, there has been a double-column fad with urban-combat roots. When moving through thickly vegetated or poorly lit terrain, such an array is unwieldy. (See Figures 13.3 and 13.4.) Only from a single file, can a patrol respond easily to any challenge. For contingency missions, the patrol leader needs only one fire team free of march security duties. (See Figure 13.5.)

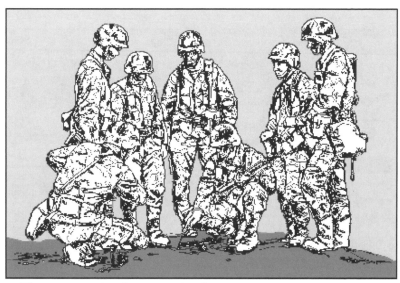

Figure 13.2: No Set Ways for Security Teams to Rejoin Line
(Source: "Army/Marine Clipart," U.S. Air University (www.au.af.mil/au/awc/awcgate/cliparmy.htm), image designator "1-07a.tif.")

Figure 13.3: Easiest Movement Formation in Close Terrain
(Source: Courtesy of Sorman Information and Media, from "Soldf: Soldaten i falt," @2001 by Forsvarsmakten and Wolfgang Bartsch, Stockholm, p. 225)

Where Contact Is Imminent

Largely administrative in nature, the above-mentioned sideways motion is directed by the patrol leader. It could be just as useful for bypassing a danger area (break in the foliage) as for investigating

141

Figure 13.4: Best Nighttime Movement Formation
(Source: FM 7-8 [1984], p. B-2)

Figure 13.5: Squad Leader Needs Only One Fire Team in Reserve
(Source: "Army/Marine Clipart," U.S. Air University, from this url: www.au.af.mil/au/awc/awcgate/cliparmy.htm)

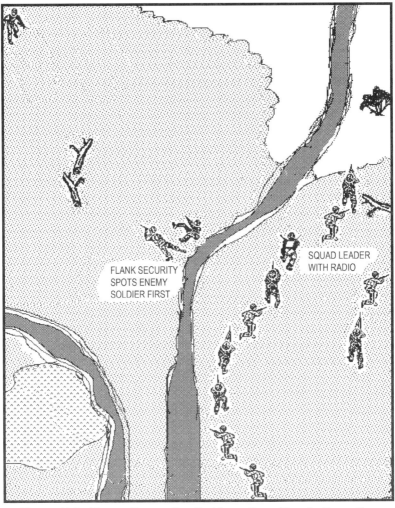

Figure 13.6: Enemy Troops Spotted from Open-Terrain Formation
(Sources: FM 7-8 [1984], p. 3-28; FM 7-70 [1986], p. 4-20; FM 7-11B1/2 [1978], p. 2-II-A-1.2; MCI 03.66a [1986], p. 2-9)

a peculiarity. Where enemy forces are definitely present, a more "tactical" variation will be required. (See Figure 13.6.) To save time, it won't be the squad leader making the decision to change forma-

Figure 13.7: All Squad Members React to Flanker's Signal
(Sources: FM 22-100 [1983], p. 185; FM 90-10-1 [1982], pp. B-3, B-4)

UPON SEEING THE SIGNAL FOR CONFIRMED ENEMY, ALL SQUAD MEMBERS GET ON LINE WITH THE SECURITY ELEMENT MAKING THE SIGHTING. THERE MUST NOW BE AN ENTIRE FIRE TEAM ON EITHER SIDE OF THAT ELEMENT.

Figure 13.8: Imminent-Contact Technique
(Sources: FM 7-8 [1984], p. 3-28; FM 7-70 [1986], p. 4-20; FM 7-11B1/2 [1978], p. 2-II-A-1.2; MCI 03.66a [1986], p. 2-9)

tion, but the security team making the sighting. Then, by forming up on this team, the skirmisher line will be automatically pointed in the right direction (e.g., 350 degrees magnetic). It will then be on autopilot, as the squad leader works up a call for fire. (See Figures 13.7 through 13.9.)

IMAGINARY TWO-MAN LANES

ALL TWO-MAN TEAMS PICK PARALLEL LANES AND MOVE FORWARD SIMULTANEOUSLY, GUIDING ON THE CENTER. IN EACH, BUDDIES TAKE TURNS STALKING AND COVERING. AS THEY DO, SQD.LDR. WORKS UP CALL FOR FIRE

Figure 13.8: Imminent-Contact Technique (continued)
(Sources: FM 7-8 [1984], p. 3-28; FM 7-70 [1986], p. 4-20; FM 7-11B1/2 [1978], p. 2-II-A-1.2; MCI 03.66a [1986], p. 2-9)

How to counter an opposition ambush would most quickly be determined by point men or flankers as well. The associated technique will appear on Table 13.2.

IF AN ENEMY FORCE STARTS SHOOTING, THE U.S. SQUAD CAN USE THE SAME LANES TO "FIRE AND MOVE" TOWARD IT. AS LONG AS ALL MEMBERS HAVE COVER AND FIRE SUPERIORITY, THEY CAN CONTINUE ON. AT THE END, THOSE CLOSEST TO THE ENEMY CAN THROW GRENADES.

Figure 13.8: Imminent-Contact Technique (continued)
(Sources: FM 7-8 [1984], p. 3-28; FM 7-70 [1986], p. 4-20; FM 7-11B1/2 [1978], p. 2-II-A-1.2; MCI 03.66a [1986], p. 2-9)

Bottom-echelon decision making can be helpful to more than just squad-sized patrols. During a large unit move to its offensive objective, the point element can better determine the final avenue

Figure 13.9: Squad Resumes Patrol from Slightly Different Spot
(Sources: FM 7-8 [1984], p. 3-28; FM 7-70 [1986], p. 4-20; FM 7-11B1/2 [1978], p. 2-II-A-1.2; MCI 03.66a [1986], p. 2-9)

Figure 13.10: Ongoing Problem with Special-Operator Inserts
(Source: "The Battle of Takur Ghar," by Keith Rocco, poss. ©, from this url: http://www.nationalguard.mil/resources/photo_gallery/heritage/Takur_Ghar_full.jpg)

of approach and point of attack (in relation to actual outposts and barriers) than the most insightful of commanders at the assembly area.

Better Squads Finally Give Big Units the Surprise Option

U.S. infantry and special operations units have seldom taken an enemy position completely by surprise over the years. There has always been some compromise to the element of surprise. Either their helicopter made too much noise coming in, or their ground approach was spotted. (See Figure 13.10.) As a result, their members never got close enough to minimize defender response.

A lead squad of more proactive and skilled riflemen would largely solve this problem. Its point buddy team could locate and bypass the listening post, cut a hole in the wire, and disarm any claymores. Once the whole squad had entered the position, it could help the rest of the unit to do likewise. At least, that's how Asian Communist armies lay the groundwork for a major attack.

Least Risky Way to Attack Is by Short-Range Infiltration

Because U.S. forces have traditionally attacked *en masse* during the day, most Americans think a few nighttime sappers have no chance. That couldn't be further from the truth. Where the quarry's fields of fire have been partially cleared or sentries put on reduced alert, sappers can sneak in at noon. If Asian troops can do this, so can Americans. The former have no more patience. To develop real patience, they would have to spend time at the bottom of a Western bureaucracy. Below is an extremely rare account of how such an attack feels to a participant.

Once [in the Mekong Delta in 1965] I got into the middle of Cai Be [ARVN] post where the district chief's office was. It was fifteen days prior to the attack and takeover of the post by our battalion. I was accompanied by two comrades armed with submachine guns to protect me in case my presence was discovered while I was nearing the post entrance. I was then wearing pants only and had in my belt a pair of pincers, a knife, and a grenade. At one hundred meters from the post I started crawling and quietly approached the post entrance with the two comrades following me. At twenty meters from the post, my comrades halted while I crawled on. At the post entrance there was a barbed-wire barricade on which hung two grenades. Behind the barricade stood a guard. I made my way between the barricade and the stakes holding up the barbed wire fence. I waited in the dark for the moment when the guard lit his cigarette. I passed two meters away from him and sneaked through the entrance. On that occasion I was unable to find out where the munitions depot was but I did discover the positions of two machine guns and the radio room. I got out at the back of the post by cutting my way through the barbed wire.[12]

— Reconnaissance sapper
261st VC Battalion

That short-range infiltration is safer than other types of attack is hard for desk-bound leaders to see, so next is offered a pictorial account of what a pair of U.S. infantrymen might experience during a squad-sized version. (See Figures 13.11 through 13.62.)

Figure 13.11: Attack Objective from a Distance
(Source: FM 7-8 [1984], p. 69)

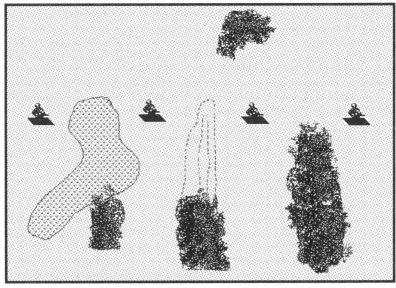

Figure 13.12: What Enemy Lines Look Like
(Sources: FM 7-11B3 [1976], p. 2-VII-C-4.4; FM 7-70 [1986], p. 4-20; FM 5-103 [1985], p. 4-6)

Figure 13.13: Single Position during the Day
(Source: FM 7-8 [1984], p. B-1)

Figure 13.14: Same Treeline at Night
(Source: FM 7-8 [1984], p. B-3)

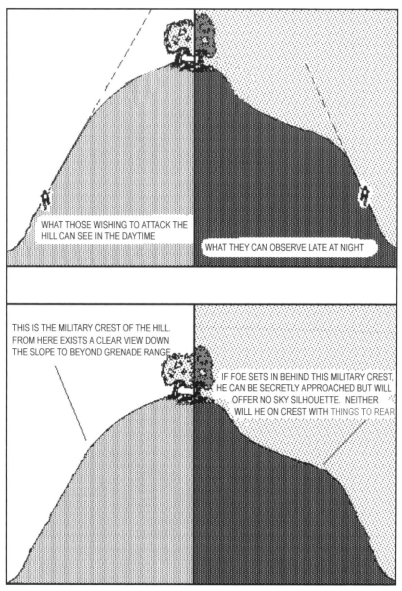

Figure 13.15: Military-Crest Observation Issues

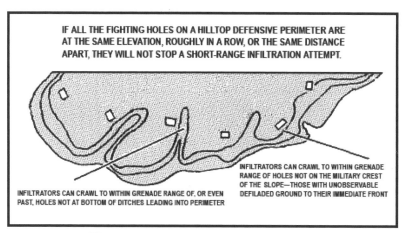

Figure 13.16: Microterrain Properties

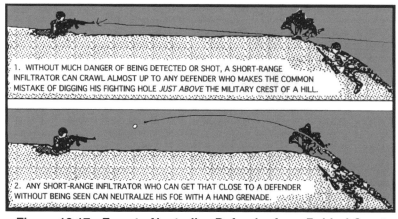

Figure 13.17: Easy to Neutralize Defender from Behind Crest
(Sources: FM 7-8 [1984], p. 3-28; FM 7-11B1/2 [1978], p. 2-III-E-8.4; FM 7-11B3 [1976], p. 2-VII-C-4.4)

Figure 13.18: Short-Range-Infiltration Lane Possibilities
(Sources: FM 23-30 [1988], p. 2-8; FM 90-5 [1982], p. 1-5; FM 90-3 [1977], p. 4-9; FM 5-103 [1985], p. 4-4; FM 7-8 [1984], p. 69)

Figure 13.19: At Start of Lane with Pre-Reconnoitered Gap
(Source: FM 7-11B1/2 [1978], p. 2-II-B-3.3)

SUCH SQUADS TAKE RAIDER-LIKE TRAINING

Figure 13.20: Definitely a Two-Man Job
(Source: FM 22-100 [1983], p. 36)

Figure 13.21: Some Gear Adjustments in Order
(Source: Courtesy of Cassell PLC, from "Uniforms of the Elite Forces," ©1982 by Blandford Press Ltd., Plate 1, No. 2)

Figure 13.22: Lane Partners Move Forward
(Source: MCRP 3-02H [1999], fig. I-2)

Figure 13.23: There Are Things to Look For along the Way
(Source: "Podgotovka Razvegchika: Sistema Spetsnaza GRU," © 1998 by A.E.Taras and F.D. Zaruz, p. 173)

Figure 13.24: Enemy Footprints
(Source: Courtesy of Paladin Press, from "Tactical Tracking Operations," © 1998 by David Scott-Donelan, p. 42)

Figure 13.25: Telltale Prongs of Anti-Personnel Device
(Source: FM 21-76 [1957], p. 38)

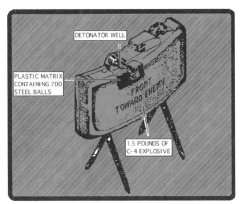

Figure 13.26: Other Types of Mines
(Source: FM 7-11B1/2 [1978], p. 2-IV-B-1.2)

159

Figure 13.27: Trip Flares
(Source: Courtesy of Sorman Information and Media, from "Soldf: Soldaten i falt," © 2001 by Forsvarsmakten and Wolfgang Bartsch, Stockholm, p. 370)

Figure 13.28: A Sudden Bright Light May at Any Time Occur
(Source: FM 7-8 [1984], p. 3-57)

Figure 13.29: Crawling through Lowest Ground Offers Benefits
(Source: MCO P1500.44B, p. 12-63)

Figure 13.30: Up-Close, Many Mounds Provide Sky Silhouettes
(Source: FM 19-95B/CM [1978], cover; MCRP 3-02H [1999], fig. I-3)

Figure 13.31: Every Sentry Has Open Areas to His Front
(Source: FM 7-11B1/2 [1978], p. 2-II-A-4.2)

Figure 13.32: Only Way to Cross Such an Area
(Source: FM 21-75 [1967], p. 67)

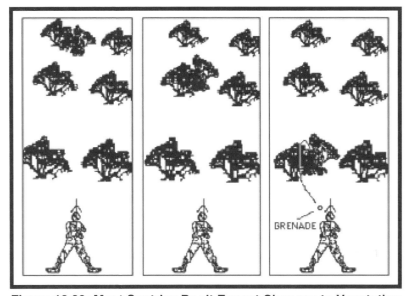

Figure 13.33: Most Sentries Don't Expect Changes to Vegetation
(Source: FM 7-11B1/2 [1978], p. 2-II-E-8.2; FM 7-8 [1984], p. 3-28)

Figure 13.34: Damp Areas Are Seldom Defender Occupied
(Source: FM 21-76 [1970], p. 43)

163

THE SURFACE OF THE EARTH IS COVERED WITH TINY WATERSHED DITCHES. BECAUSE MOST DEFENSES ARE ON ELEVATED GROUND, THEY HAVE DITCHES LEADING INTO THEIR INTERIOR.

Figure 13.35: Watershed Ditches Run from All Elevated Ground
(Source: FM 5-103 [1985], p. 4-4)

Figure 13.36: Some Fields of Fire Can't Be Fully Cleared
(Source: FM 100-20 [1981], p. 178)

Figure 13.37: Those from the City Might Try to Hide This Way
(Source: OPNAV P 34-03 [Revised 1960], p. 394)

Figure 13.38: Country Hunters Know Better
(Source: FM 6-13E1/2 [1979], p. 2-359)

165

Figure 13.39: Moving through Heavy Brush Makes Noise
(Source: "Podgotovka Razvegchika: Sistema Spetsnaza GRU," © 1998 by A.E.Taras and F.D. Zaruz, pp. 144, 145, 152)

Figure 13.40: Rain Helps to Muffle That Noise
(Source: Corel Gallery Clipart, Weather, #46A004)

Figure 13.41: So Does a Heavy Wind
(Source: Corel Gallery Clipart, Weather, #46A004)

Figure 13.42: It's Best Masked by Distant Shelling or Machinery
(Sources: FM 7-8 [1984], p. 4-25, MCO P1500.44B, p. 15-26; MCI 7311B, p. 115)

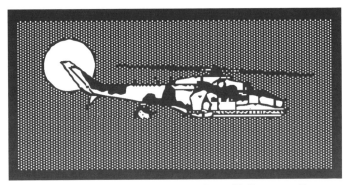

Figure 13.43: Much Is Possible As a Helicopter Passes
(Sources: FM 7-8 [1984], p. 4-25; MCO P1500.44B, p. 15-26; MCI 7311B, p. 115)

Figure 13.44: Fallen Log Offers Line Passage Opportunity
(Sources: FM 7-11B1/2 [1978], p. 2-II-A-8.2; FM 7-11B3 [1976], p. 2-VII-C-4.4)

Figure 13.45: With Any Moon, Trees Throw Shadows
(Source: MCRP 3-02H [1999], fig. I-5)

Figure 13.46: Very Slow Movement Imperceptible in Low Light
(Sources: FMFM 0-1 [1979], p. 7-7; FM 21-75 [1967] p. 2-II-A-4.2)

Figure 13.47: Watchstanders Doze Late at Night
(Sources: FM 23-30 [1988], p. 2-8; FM 90-5 [1982], p. 1-5; FM 90-3 [1977], p. 4-9; FM 5-103 [1985], p. 4-4; FM 7-8 [1984], p. 69)

Figure 13.48: Stalker Keeps Object between Himself and Quarry
(Source: "Podgotovka Razvegchika: Sistema Spetsnaza GRU," © 1998 by A.E. Taras and F.D. Zanuz, p. 373)

Figure 13.49: Any Low Mound or Rock Will Work
(Source: Courtesy of Sorman Information and Media, illustration by Wolfgang Bartsch, from "SoldF: Soldaten i falt," © 2001 Forsvarsmakten, Stockholm, p. 220)

Figure 13.50: Sentry Can't See Right After Cloud Covers Moon
(Source: FM 7-11B1/2 [1978], p. 2-II-A-4.2)

Figure 13.51: Distant Flare Makes Night Vision Device Useless
(Sources: FM 7-11B1/2 [1978], pp. 2-11-A-4.2 , 3-11-A-4.2; OPNAV P34-03 [1960], p. 40)

Figure 13.52: Even Smoking Keeps Defender from Seeing Much
(Sources: FM 7-11B1/2 [1978], pp. 2-11-A-4.2 , 3-11-A-4.2; OPNAV P34-03 [1960], p. 40)

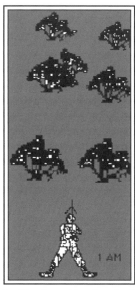

Figure 13.53: Separate Bushes Harder to Distinguish after Dark
(Sources: FM 7-11B1/2 [1978], p. 2-II-E-8.2; FM 7-8 [1984], p. 3-28)

Figure 13.54: Hole Occupant Distracted by Line Checker
(Sources: FM 100-5 [1994], p. 39; FM 5-103 [1985], p. 4-5)

Figure 13.55: Watch Change Grabs Everyone's Attention
(Source: FM 90-5 [1982], p. 2-12; FM 100-5 [1994], p. 37)

Figure 13.56: Elsewhere Noises Preoccupy Sentries
(Source: FM 20-32 [1976], p. 151)

Figure 13.57: Other-Lane Infiltrator May Create a Diversion
(Source: MCRP 3-02H [1999], fig. I-2)

Figure 13.58: Both Sappers Go to Where They Can See Sentries
(Source: FM 5-103 [1985], pp. 5-2, D-9)

175

Figure 13.59: Lead Man Visualizes Move after Studying Gap
(Source: "Podgotovka Razvegchika: Sistema Spetsnaza GRU," © 1998 by A.E. Taras and F.D. Zanuz, p. 373)

Figure 13.60: Penetration Attempt after Some Diversion Occurs
(Sources: FMFM1-3B [1981], p. 4-9; FM 5-103 [1985], p. 4-4)

Figure 13.61: Backup Man Keeps Bead on Closest Sentry's Head
(Sources: FM 7-11B/C/CM [1979], p. 2-5; FM 5-20 [1968], p. 31)

Figure 13.62: Should That Sentry Get Nervous
(Source: FM 17-1 [1966], p. 179)

177

As visualization and daring both became necessary to accomplish this feat, the two GIs involved would have welcomed a little "assertiveness" training. No American special operator should attempt this type of thing without more light-infantry experience.

What Makes This Type of Attack So Safe

Instead of forcing one's way into the enemy camp, the above progression follows an Oriental axiom: "Wait for nature and other circumstances to provide an opportunity." That's what Figures 13.50 through 13.57 were all about. By reaching the assigned gap around 1:00 A.M., each team will have a couple of hours to study it and capitalize on any natural break to the sentries' vision or concentration. The two hours of darkness left should be plenty of time for rendezvous, explosive rigging, and exfiltration.

As the two team members take turns passing through enemy lines, each draws a bead on the nearest sentry's head. That sentry is not be bothered, if possible. His demise would forfeit a well-rehearsed exit route. After sneaking by him from the front, passing from the rear on the way out should be easy. Upon completion of the interior mission, there's less of a reason to keep that sentry conscious. But, with no evidence of frontline entry, the compound commander is more likely to chalk up all inside damage to an accidental occurrence or lucky mortar hit. Then, the same technique will work just as well the next night.

Every element of such a short-range-infiltration attack can complete the mission alone. If a tank, command bunker, or ammunition dump is to be blown, each team carries enough explosive and a timer (to permit pre-detonation escape). That means fewer people ultimately jeopardized. In the attack just portrayed, only one of the squad's seven teams was obligated to sneak between enemy holes. None of the others (after assessing their chances from close range) had even to try. Their only absolute requirement was to grenade the enemy holes to their front should an emergency escape route become necessary.

Slightly over a dozen GIs took part in the above attack, and only the most confident pair were to take much risk in its execution. That's considerably safer than hundreds of partially trained Americans dutifully rushing the dug-in machineguns that will invariably survive any preparatory bombardment.

Other Squad Capabilities

When finally comprised of more proactive fighters, U.S. infantry squads will be capable of all kinds of amazing maneuvers. Many are things their German, Japanese, Chinese, North Korean, and North Vietnamese counterparts have already proven to work. For those who don't follow the evolution of squad tactics, others may seem a bit fanciful. Still, all are viable, and they will permit a much broader range of large-unit tactics. Now available are even the specifics of each "breakthrough" technique. (See Figure 13.63 and Table 13.2.)

Figure 13.63: Squads Must Disperse to "Control Neighborhoods"
(Source: public domain material from U.S. Army "War on Terror Images," at this url: http://www.history.army.mil/books/wot_artwork/images/35a.jpg)

CONVENTIONAL PATROLLING EXPERTISE

SAFELY AMBUSHING ANY NUMBER OF ENEMY	LHYCHAPT16
COUNTERAMBUSHING ANY NUMBER OF ENEMY	LHYCHAPT15
BETTER (DECENTRALIZED) SEARCH PATTERN	EECHAPT14

CONVENTIONAL APPROACH EXPERTISE

SECRETLY MOVING IN ON RURAL OBJECTIVE	LHYCHAPT17,DDCHAPT15
	DDCHAPT17(PP284-285)
SECRETLY MOVING IN ON URBAN OBJECTIVE	DDCHAPT16(PP244-248)

CONVENTIONAL ATTACK EXPERTISE

SHORT-RANGE INFILTRATION ATTACK	LHYCHAPT20,GWCHAPT17
	DDCHAPT15(PP230-235)
DAY ATTACK BY COMPLETE SURPRISE	LHYCHAPT18,GWCHAPT17
REGULAR NIGHT ATTACK BY SURPRISE	LHYCHAPT19,GWCHAPT17
INDIAN FILE NIGHT ATTACK	TJCHAPT14(PP207-213)
RAIDING A STRATEGIC SPOT BEHIND ENEMY LINES	DDCHAPT17(PP257-279)
MORE DECISIVE AND SAFE BUILDING ASSAULT	HSCHAPT9
ATTACKING ALONG STREET AND AT INTERSECTION	LHYCHAPT24
TAKING BUNKER BELT WITHOUT SUPPORTING ARMS	GHCHAPT10(PP127-131)

CONVENTIONAL DEFENSE EXPERTISE

REARWARD-MOVING URBAN STRONGPOINT	LHYCHAPT23
DENYING RURAL OUTPOST TO ENEMY HORDE	EECHAPT15

UNCONVENTIONAL WARFARE KNOWLEDGE

OPERATING A SQUAD ZONE IN ENEMY COUNTRY	DDCHAPT19(PP311-320)
COVERING AREA WITH SECRET FIRE TEAM ZONES	TJCHAPT14(PP194-206)
BREAKING OUT OF RURAL ENCIRCLEMENT	DDCHAPT18(PP289-301)
BREAKING OUT OF URBAN ENCIRCLEMENT	DDCHAPT20(PP332-344)
	DDCHAPT21(PP358-362)
HIDING FROM ANNIHILATION IN THE COUNTRY	DDCHAPT19(PP321-325)
HIDING FROM ANNIHILATION IN THE CITY	DDCHAPT21(PP350-357)
CONTROLLING A NEIGHBORHOOD	MTCHAPT13(PP248-256)
	TJCHAPT16

CODE:

LHY, THE LAST HUNDRED YARDS, ISBN ISBN 0963869523
DD, DRAGON DAYS, ISBN 096386954X
MT, MILITANT TRICKS, ISBN 0963869582
TJ, TEQUILA JUNCTION, ISBN 0963869515
HS, HOMELAND SIEGE, ISBN 9780981865911
EE, EXPEDITIONARY EAGLES, ISBN 9780981865928
GW, GLOBAL WARRIOR, ISBN 9780981865935
GH, GUNG HO, ISBN 9780981865942

Table 13.2: Semi-Autonomous Squad Capabilities

On this Table, please note all the surprise options these techniques now make possible for large U.S. units (by using a single squad as offensive spearpoint or defensive building block). Within the UW portion is how this same unit could blanket a huge area with its limited manpower. (Look again at Figure 13.63.) Of course, more opportunistic squads will require commensurate restrictions on their behavior. (See Figure 13.64.)

Figure 13.64: Level of Force Harder to Manage for Dwelling Entry
(Source: "Capture at Ar Ramadi," by Don Stivers, poss. ©, from this url: http://www.nationalguard.mil/resources/photo_gallery/heritage/hires/arramadi.jpg)

How Useful Is This New Information?

No scheme, however beneficial to national security, will succeed without enough supporting details. The final part provides them to a degree that most U.S. infantry officers have not felt necessary. But then, war has changed. Semi-autonomous enlisted contingents will play a bigger role in the next one.

Part Five

Properly Preparing the New Squad Member

"INFANTRY IS . . . STILL THE DECISIVE FACTOR IN COMBAT.
FOR IT IS THE FOOT SOLDIER WHO TRAVELS THE LAST ONE HUNDRED YARDS
TO A DECISION. . . . [A]S THE INFANTRYMAN APPROACHES THE ENEMY LINES,
ALL . . . SUPPORTING FIRE MUST BE LIFTED. HE IS "ON HIS OWN." . . .
THEN THE OUTCOME RESTS ENTIRELY ON THE EFFECTIVENESS OF HIS OWN
INDIVIDUAL WEAPONS—THE RIFLE, THE BAYONET, . . . THE HAND GRENADE.
AND MOST OF ALL, ON THE DOUGHBOY'S COURAGE AND SKILL."
— WWII WAR BOND AD

(Source: 1944 Sperry Corporation War Bond Ad #A5514 [07/17/44])

14 All-Hands Accountability for More Moral Units

- ● Which rules of conduct won't suppress rifleman initiative?
- ● Why are Privates a good monitor of unit morality?

Maneuver-savy fighters better handle behavioral restrictions.

(Source: public-domain material from U.S. Army "War on Terror Images," at this url: http://www.history.army.mil/books/wot_artwork/images/47a.jpg)

Wartime Riflemen Already of High Moral Fiber

When the bullets start flying, those with no desire to be there generally aren't. That makes combat infantrymen a very special segment of the U.S. military population. For enlisted personnel, any heroism citation is generally a notch lower than that recommended—e.g., a Bronze instead of Silver Star. Just to win America's second highest award, recent recruits would have to earn the highest many times over. Yet, the number of Privates, PFC's, and Lance Corporals to have been awarded the MoH over the years is truly amazing. Only Leatherneck actions will be analyzed here, as their

Service Branch is generally thought to be the most stingy with its citations. During WWII, 26 of 75 Marine MoH recipients were of pay grades E-1 through E-3. At Korea, it was 13 out of 42. In Vietnam, it was 30 out of 57.[1] That's a full 40% of the total (spanning some 28 ranks). If this statistic doesn't adequately attest to the depth of the combat nonrate's character, nothing will.

Though the Pentagon's junior enlisted members may be a little more "ragged around the edges" than their commissioned counterparts, they are every bit as ethical in combat. After 15 months of almost continuous campaigning in Vietnam, one rifle company commander claims his troops collectively provided him with the best "moral compass." While his battalion commander once denied (unsuccessfully) a request to medivac civilian casualties, his men wouldn't even let him burn a VC hooch. During the two days out of every month his company spent re-outfitting, he occasionally ran into a "combat PFC" on the streets of a forward base. It was then that this Captain felt the irresistible urge to salute first. In his mind, there was no doubt where the heart and soul of his unit lay. Those 18-year-olds did almost all the fighting, and thus deserved most of the credit—including that for any medals their commander may have received.[2] When a former infantry leader sports such a trophy, it is normally for what his troops have accomplished (often with too little help from him).

Riflemen Can Be More Than Extensions of Their CO

After encountering 4GW opposition in Iraq and Afghanistan, the Pentagon again sees indigenous forces as the only way to win most modern wars. Such forces usually have few tanks, planes, and artillery pieces, so U.S. assistance most effectively takes the form of light-infantry instruction. Unfortunately, even America's tactical elite—its special operators—have so focused on technology over the years as now to lack the prerequisite light-infantry expertise. To develop it, they would have to return to what most Americans consider to be lowly "basics"—shooting, moving, and communicating. Far from elementary, each is actually a vast assortment of complicated subsets. For example, an important part of moving is how silently to crawl through dry leaves or swampy ground. And, as asserted by Part Five's introductory war bond ad, grenade throwing is every bit as vital to ground combat as firing one's rifle. (See Figure 14.1.) Implicit in a "Doughboy's skill" is his ability to use all available cover while approaching an enemy defender. Among other things, that takes continually choosing between various speeds and postures. Since tactical-decision making

Figure 14.1: Grenades Still Necessary to Close with the Foe
(Source: "Lend the Way," U.S. Army Center of Military History, from this url: http://www.history.army.mil//art/Posters/WWI/Lend_the_Way.jpg)

is almost never practiced by American inductees, it will first be addressed. That way it can be added to the application phase of some movement instruction.

Rules That Won't Suppress Initiative

Because the U.S. military has traditionally practiced 2GW (killing as many enemy as possible), its entry-level infantry instruction has largely focused on doing exactly as told at the lowest echelons. Theoretically, that's so the commander's intentions can be more precisely followed. But, his actual orders will seldom contain much frontline detail. And now that wars are mostly fought in a 4GW environment (with political and religious overtones), the rank and file must rely on more "common sense" in their prosecution. Too many of the wrong kinds of rules would prevent that much thinking on their part.

"Bottom-up" Asian armies partially solve this problem with troop standards that mostly address noncombatant treatment. Should any soldier then become confused as to what to do next in battle, such rules would limit behavioral excess but not tactical initiative. Below are the ones still followed by Mainland Chinese riflemen as of 1975.[3]

> Three Main Rules of Discipline:
> Obey all orders
> Do not take a . . . needle or . . . thread from the masses
> Turn in everything captured
> Eight Points of Attention:
> Speak politely
> Pay fairly for what you buy
> Return everything that you borrow
> Pay for anything that you damage
> Do not hit or swear at people
> Do not damage crops
> Do not take liberties with women
> Do not ill-treat captives [4]
> — *Handbook on the Chinese Armed Forces*

Most deserving of a closer look is "order obeying." Its meaning would differ in a "bottom-up" organization. A low-ranking participant of as much Guerrilla as there was Mobile and Positional Warfare might get more open-ended directives.[5] Such mission-type or "frag" orders would automatically give him more wiggle room in their compliance. A *Defense Intelligence Agency (DIA) Handbook* describes "the average [Chinese] soldier as . . . able to . . . improvise under a wide variety of conditions."[6] If that soldier had to obey—to the letter—every utterance of his immediate superior, he would be unable to improvise anything.

A Different Way of Maintaining Order

Also of interest is how justice was meted out in a PLA unit. Only major conduct violations were handled by courts martial. Lesser infractions were left to peer pressure. In a semi-autonomous squad, peer pressure would be absolutely vital to avoiding the atrocity that might emerge out of marginal leadership. Peer pressure is a key ingredient of unit cohesion.

> Minor offenses (like drunkenness or failure to care for equipment) are usually dealt with at company-level criticism meetings. . . . At the meeting, the soldier confesses his crime and pledges to reform. The attendees [his peers] decide on an appropriate punishment.[7]
> — *DIA Handbook on the Chinese Armed Forces*

PRC Troops May Have Had Few Rules on How to Fight

Besides safeguarding civilians, the Chinese nonrate was restricted from being "impolite or abusive" to any military leader, contemporary, or captive. Yet, nowhere in his code of conduct is there any mention of how to deal with an immediate adversary. Whether his parent unit had any Rules of Engagement that would have further restricted his actions is not clear. Still, this apparent lack of interference with how each soldier would fight made it easier for him to exercise a little tactical judgment.

Though enjoying more of a collective "say-so" than most Western counterparts, PLA troops were also more afraid of their leaders. That almost certainly limited any initiative on their part to a few well-established categories.

189

Each PLA Soldier Was to Monitor Parent-Unit Progress

Americans have come to better understand the Asian Communists' "bottom-up" way of running a unit (leadership by consensus) through the *"Gung Ho* Sessions" that Carlson permitted his Raiders.[8] Like the *"Kiem Thao* Sessions" with which the North Vietnamese did so well in Vietnam,[9] these group gatherings assisted in both after-action reporting and combat deficiency correction. If the Chinese Reds or Marine Raiders had been given an additional "ethics-monitoring" assignment, their parent units might have less easily strayed from their own moral intent.

According to the movie classic *"Gung Ho,"* all Raiders were briefed on the battalion's plans and objectives before every operation, and then asked to critique its success in open forum afterwards.[10] In other words, each low-ranking Marine was to be an inspector of unit progress. If such progress had been defined as less collateral damage, that's what each Private would have championed.

With so "civilian-friendly" a code of conduct, Chinese forces would have been perfectly suited for widely blanketing an area with tiny, semi-autonomous detachments. Isn't that what U.S. units have lately attempted (with varying degrees of success) in Iraq, Afghanistan, and Africa? With a similar set of rules, they might avoid the occasional embarrassment that has so far prevented enough decentralization of control.

U.S. Riflemen Could Help to Achieve Moral Objectives

When the North Koreans invaded South Korea and Chinese moved into the Chosin Reservoir region, they both did so as "veritable phantoms."[11] They employed very little firepower. It can thus be said that they were practicing 3rd-Generation Warfare (3GW)—where troop concentrations are intentionally bypassed to more easily get at a foe's strategic assets.

This more advanced way of fighting over 60 years ago should come as no surprise, because Maoist Mobile Warfare closely resembles contemporary MW.[12]

Fighting in such a way as not to need much firepower would certainly limit most forms of collateral damage. However, Com-

Figure 14.2: There Can Be No Fraternization with Local Women
(Source: http:// search.usa.gov public-domain image from this url: http://www.history.navy.mil/pics/ng-16.jpg)

munist commanders did regularly hang local mayors in Korea. So every public-relations standard would have to be universally followed by all ranks.

As U.S. Marines come to more fully embrace their new MW doctrine (finally incorporate advanced squad tactics), their operations should require less brute force. Like the Raiders for other command goals, contemporary riflemen could then monitor parent-unit progress on a wide assortment of moral issues. (See Figure 14.2.)

Figure 14.3: After Foe Troop Concentration Has Been Bombarded
(Source: "Jungleers on Biak," by Keith Rocco, poss. ©, from this url: http://www.nationalguard.mil/resources/photo_gallery/heritage/hires/Jungleers_on_Biak.jpg)

Of course, where killing remains the GI's primary focus (as in a "higher-tech" version of 2GW), any decrease in collateral damage would be hard for him to identify. (See Figure 14.3.)

15 Reestablishing _ Individual Initiative

- Are U.S. recruits punished for inventiveness?
- How might such a quality be militarily refocused?

That all-American trait can be quite helpful in close combat.

(Source: "Under Fire," by SFC Darrold Peters, from U.S. Army "War on Terror Images," at this url: http://www.history.army.mil/books/wot_artwork/images/52a.jpg)

Bravery and Initiative Are Connected

Heroism is generally defined as continuing on with something of which one is greatly afraid. Reckless or disconsolate behavior (like the Lieutenant's "leadership by example" in Chapter 7) doesn't qualify. Depending on the degree of peer pressure within a unit, uncommon valor may become a common virtue. Anyone who has studied the 36-day battle for Iwo Jima can attest to every frontline Marine's repeated display of enough courage to win a medal. The same can be said of those who fought for the Sugar Loaf Complex on Okinawa, or any of the U.S. Army's more desperate battles.

Though peer pressure can create a more universal display of heroism in a unit, it can also lead to unnecessary casualties. Marine veterans of the Korean War often tell of infantrymen feeling so guilty about one day's fully justified caution as to go out and get themselves killed the next. Much of the male "stay-at-home" population of England felt the same way about not attending the WWI slaughter. So an interesting question is posed. How might enlisted courage be most effectively utilized in war?

There's More Than One Way to Display Bravery

A closer look at the MoH citations for E-1 through E-3 from WWII forward reveals a preponderance of "grenade jumping" to save buddies. Though many wartime acts are of equal lethality and longer duration, their details are harder for headquarters to establish in the heat of an ongoing campaign. Many fatalities occur as frontline troops are doing more than their duty. Either no one sees them do it, or their contribution gets lost among all the others. Then, their only reward will be the standard letter to parents about being killed in action and a Purple Heart with black edges. Such is the lot of untold numbers of blessed American war dead—each with his (or her) own secret story of how exceptionally well they performed on the last day before going to meet the angels.

Within the U.S. military, every young rifleman's dream is some heroic deed on the field of honor. To keep casualty figures low, that ever-so-natural impulse must be carefully channeled. One former battlefield commander puts it this way: "Marines are hard to keep alive because they're so damn brave."[1] He and his current contemporaries would love to have some way of more safely expending all that potential. Well, should any U.S. commander now be interested, the Communist Asian armies have long enjoyed such a way. And one doesn't have to be Communist to use it.

Must All Valor Be on the Spur of the Moment?

American tactics "aficionados" now readily admit to U.S. ground forces seldom being able to surprise enemy defenders from short range. Their attacks have been mostly of the "high-diddle-diddle-up-the-middle" variety, with as much firepower as the assault force can muster. Sometimes it works, and sometimes it doesn't. When it doesn't, that force gets pinned down. Then, someone has to come up with enough "uncommon valor" to silence the offending machinegun. In other words, the original (unimaginative) maneuver must be supplanted with an impromptu breakout. Such a task usually falls to the nearest NCO and some nonrates. Still, one wonders how their bravery might have been put to better use. What if the original maneuver had involved a little surprise? Then, there wouldn't have been as much need for all that last-minute, death-defying exposure.

Why Not Incorporate More Clever Small-Unit Maneuvers?

Danger in combat is a relative term. Assaulting hidden enemy machineguns *en masse* while fully upright and on line is not very life-enhancing behavior. Yet, commanders wince at the thought of a few men trying—while prone—to sneak through enemy lines. Chapter 13 has already demonstrated how such a thing can be quite safely done. Though far less risky than running through streams of machinegun bullets, this Asian way of conducting an assault is almost never attempted by U.S. forces. It does require some participant courage, but at least that courage is not being expended on a makeshift and often futile try to escape an active firesack. Only one of several short-range infiltration teams has to get through for the unit's mission to be accomplished. This gives any team uncomfortable with its chances, the option to abort. Just working this close to enemy sentries would be enough to sustain their morale.

Who Wins Most Battles?

To what is the success of an American ground attack mostly owed? Is it the weapons proficiency, tactical wisdom, and charismatic leadership of its unit commander, or the combined contributions of his troops? After the assault force has been detected, aren't those ordinary U.S. riflemen up and running through withering enemy fire?

Of only three MoH's for the Omaha Beach landing, one went to Private Carlton W. Barrett.[2] There must have been hundreds of acts of extreme heroism on that terrible day. Why would U.S. authorities single out a Private who actually survived it? It was at Omaha Beach that badly wounded GIs were to crawl up to the deepest penetration to protect others from (and personally absorb) incoming bullets. So, the answer must lie in the fact that Private Barrett's initiative had helped to make the breakout possible. Like Barrett, the other enlisted MoH winner—Technician 5th Grade John J. Pinder, Jr.—had restored communications between units.[3] Such a conclusion is not often drawn, because Private Barrett first kept several people from drowning. (See Figures 15.1 through 15.3.)

> For gallantry and intrepidity at the risk of his life above and beyond the call of duty on 6 June 1944, in the vicinity of St. Laurent-sur-Mer, France. On the morning of D-day, Pvt. Barrett, landing in the face of extremely heavy enemy

Figure 15.1: Enemy Fire Starting Coming In Hot and Heavy

Figure 15.2: Many Never Reached the Beach
(Source: http://search.usa.gov public-domain image from url: www.nps.gov/history/history/online_books/npswapa/extContent/usmc/pcn-190-003120-00/images/fig65.jpg)

Figure 15.3: Private Barrett First Tried to Drag In the Floundering
(Source: http://search.usa.gov public-domain image from url: www.nps.gov/history/history/online_books/npswapa/extContent/usmc/pcn-190-003120-00/images/fig1.jpg)

fire, was forced to wade ashore through neck-deep water. Disregarding the personal danger, he returned to the surf again and again to assist his floundering comrades and save them from drowning. Refusing to remain pinned down by the intense barrage of small-arms and mortar fire poured at the landing points, Pvt. Barrett, working with fierce determination, saved many lives by carrying casualties to an evacuation boat lying offshore. In addition to his assigned mission as guide, *he carried dispatches the length of the fire-swept beach;* he assisted the wounded; he calmed the shocked; he arose as a leader in the stress of the occasion. His coolness and his dauntless daring courage while constantly risking his life during a period of many hours had an inestimable effect on his comrades and is in keeping with the highest traditions of the U.S. Army.[4] [Italics added.]
— MOH Citation for Private Carlton W. Barrett

That was long ago. More recently, it was probably a junior special operator who spotted the tiny piece of rope that led to the hidden cover of Saddam Hussein's spider hole in the mud-hut compound near Tikrit.[5] (See Figure 15.4.) Was low-ranking initiative of strategic value then?

Of course, not every Private can make that big a difference to an overall war effort. Yet, it is still nonrates who are mostly responsible for all those little victories that go into winning any battle. To get the most out of modern-day "gladiators," one must encourage their inventiveness.

With proper training, our Americans are better [than the Japanese], as our people can think better as individuals. Encourage your individuals and bring them out.[6]
— Col. Merritt A. Edson, after Guadalcanal

Or is the Marine Corps' long-standing and heritage-building *mantra* of "improvise, adapt, and overcome" only restricted to its leaders?

Where Did That Uniquely American Characteristic Go?

Most freedom-loving Americans develop a fairly independent

Figure 15.4: Junior Special Operators Would Think Smaller
(Source: public-domain material from U.S. Army "War on Terror Images," at this url: http://www.history.army.mil/books/wot_artwork/images/24a.jpg)

spirit at a very early age. At the highest levels of government, such a spirit can easily be construed as "dissatisfaction" or "bucking the system." Yet, at the lower echelons of a highly stratified military bureaucracy, it may be the only way to get anything accomplished in a timely manner.

For these and other reasons, Congress has never really trusted the U.S. military. Most Armed Forces' headquarters have, in turn, placed far too much emphasis on their own directives. Combine this systemic shortage of latitude with a warfare style (2GW) that requires every Private to do precisely as told, and one has a major loss of initiative where it might do the most good. In fact, American recruits have this immensely valuable quality all but driven out of them by the time they leave boot camp. As late as 2002, their German counterparts (descendents of the highly proficient WWI Stormtroopers) were instead being punished for not showing enough initiative.[7]

To be sure, troops from group-oriented German and Oriental societies might need more of a push toward independent action. Asians, in particular, are often accused of too easily following the crowd. But then, most Yanks could be of more help to their group. In the average U.S. unit, too much enlisted leeway would lead to serious control problems. Thus, for each new American grunt, the

Figure 15.5: The Highly Responsive U.S. Marine Raider
(Source: Courtesy of Edward Molina © 2012)

initiative lost at boot camp must be only partially restored at subsequent commands. The safest way to do that is through a short list of procedures that—at the appropriate time—may be openly questioned by any fighter.

A Good Example from History

What Lt.Col. Carlson really wanted in 1942 was Raiders who were more assertive than the average Marine. To get them, he largely depended on hand-to-hand combat. Instead of just showing everyone how to fight at close quarters, he attempted to regenerate and then focus the initiative that had been left in boot camp. Each Raider was asked to watch for an attack against his person from any direction, and then—through a feint or deception of his own design—to gain the upper hand.[8] By allowing each man to choose/develop his own ruse, Carlson triggered a more opportunistic mindset.

Then, to achieve a more universal display of initiative, the 2nd Raider Battalion Commander needed ways to supplant the steady diet of orders. First, there were the implicit guidelines of his "Ethical Indoctrination." Then, there were the collective conclusions of his *"Gung Ho* Sessions." It was as the latter that nonrates could find professional fault with each other and superiors. Carlson considered this chance to publicly voice one's opinion a further way to develop a willingness, resourcefulness, and initiative among his lowest-ranking personnel.[9]

The good Colonel's new "leadership-by-consensus" model depended on everyone fully cooperating (working together in harmony). Soon, displays of initiative were regularly taking the place of unquestioning compliance at the lowest echelons of his soon-to-be famous 2nd Raider Battalion. (See Figure 15.5.)

After the Raiders were disbanded in 1943, the term *Gung Ho* first came to mean "spirit" and "can-do attitude."[10] Within the second capability was enough confidence to fend for oneself. That's nearly the same as initiative, and—within densely populated portions of Asia—a common characteristic.[11] The "bottom-up" Eastern way of war then brokers this way of upstaging the masses into more frontline-fighter initiative than its "top-down" Western counterparts can muster.

Within this paradox lies the solution to the Pentagon's chronic deficiency. Still, there are pitfalls. For the average American unit, too much lower-echelon innovation would easily erode discipline. As such, it would have to be carefully controlled. The frontline rifleman most usefully contributes to his unit's tactical and moral objectives. That's why Mao had pushed *"Gung Ho* Sessions" and "Ethical Indoctrinations."[12]

If American Raiders could learn this much about instilling the right kind of initiative from Chinese infantrymen, then there is no telling what they might discover from early Japanese commandos. Commandos more often work alone.

Secret to the *Ninja's* Success

Though the black-leotard-clad *ninja* has been the butt of many American jokes over the years, he was still present on Southeast Asian battlefields as late as 1968.[13] His extensive *repertoire* of fortress entry and exit procedures has long been the model for Asian "special operators."[14] Those who are Communist regularly train their country's light-infantry establishment, so some of these secretive techniques have undoubtedly reached Asian riflemen. Within the *ninjas'* training regimen are also ways for those riflemen to develop more initiative.

The practitioner of advanced Japanese *ninjutsu* (and its various Mainland equivalents) enjoys "integrated mind-body awareness."[15] In strict compliance with the laws of the universe, he tries not to reach an accelerated state of aggression, but rather to avoid confrontation. As such, he is at one with nature and unafraid of any antagonist. He can easily sense that antagonist's presence, ascertain his intentions, and even dodge his first unseen blow from behind. Such a fighter is convinced, and rightfully so, that he can single-handedly defeat many times his number of violence-bent pursuers.

How does this *ninja* attain so advanced a state of readiness? Under real-life conditions, he regularly rehearses any number of sensory-enhancement, individual-movement, and personal-concealment techniques. He spends long hours in the woods learning their every vibration and nuance. All the while, he is building sensory and muscle "memory." Then, by training his mind not to alter what his senses have just told him, he can do what needs to be done. He maintains this degree of readiness by working over and over on mind-body coordination.

To mesh with ever-changing and often chaotic circumstances, the *ninja* must be able to transcend technique.[16] By making many of his actions almost instinctive, he more easily focuses on his immediate adversary. By continually refining those actions (choosing between maneuver options), he gets tactical-decision-making practice. Might a similar routine be of any help to lower-echelon special operators and light infantrymen from America?

Even Mao Realized the Need for Infantryman Values

PLA units of every size must be able to shift quickly between Mobile, Positional, and Guerrilla Warfare. The latter is what gives Chinese fighters so much practice with personal initiative. Though Mao would only grant mercy to enemy soldiers as a psychological ploy, he still understood why each of his own fighters needed a "conscience." A more Western connotation of that term might be of assistance to U.S. troops in a modern, spiritually oriented war.

[T]he basis for guerrilla discipline must be individual conscience. With guerrillas, a discipline of compulsion is ineffective.[17]
— Mao Tse-Tung quote
1941 *Marine Corps Gazette*

More Troop Initiative through Guerrilla Warfare Training

Being able to think on one's feet inside a friendly formation is one thing. Arriving at logical conclusions when alone and totally surrounded by adversaries is harder. That's why any increase in personal initiative must necessarily be combined with more time away from the parent-unit umbrella.

Carlson was able to practice Maoist Mobile Warfare (a close facsimile of German MW) at the squad level on Guadalcanal, because he had trained his people to fight like guerrillas. At the Marine Raider Training Center in Southern California, many guerrilla warfare classes had been added to the curriculum. Free-wheeling aggressor duty also took the place of "canned" exercise participation.[18] All Raider trainees regularly rotated between conventional and guerrilla roles in these mock battles.

Tactical "Management by Exception" Also Works

To reestablish the right kind of individual initiative within U.S. infantry squads, their members would have to be less strictly controlled during maneuver practice. There are ways. The most obvious is for each squad thoroughly to rehearse the solution to some combat scenario, and then have its leader issue no further instructions during its execution.

Infantry squads are like football teams. To move down a field, a dozen or so people need no redirection or supervision. In fact, after enough situational guidance and technique rehearsal, they don't need any leader at all. The chance contact drill provides a perfect example. Its "fire-and-movement" evolution involves tiny teams alternately rushing along parallel lanes. Who rushes and when needs no on-site coordination or previous choreography. Once its participants realize how to stay within their lanes and roughly on line, it just happens. Anything else would be too predictable in actual combat.

There are, of course, "situational" limitations to this type of maneuver. Stand-up rushes won't work against barbed wire, claymores, or well-directed machinegun fire (the trappings of a deliberate defense). Should a free-wheeling squad spot fresh dirt, it should stop immediately. Nor can it move across terrain that is totally devoid of cover. As long as the squad leader explains these things to his men ahead of time, they can "fire and move" without him. That gives him more time to come up with the right grid coordinates for a danger-close artillery or mortar mission. Only, when his men are forced—by their initial guidelines—to stop, must the squad leader reassume control. Until then, his tiny unit is on autopilot, and its leader "managing by exception." This will give him momentum against a world-class opponent.

As Lt.Col. Carlson discovered, it takes more fighter involvement than is customary within U.S. line units to develop sufficient frontline initiative. In the above drill, that level of involvement becomes necessary. For it is each participant's initiative that will ultimately keep him from getting shot. (See Figure 15.6.) Only

Figure 15.6: Lone Rifleman Must Outsmart Immediate Adversary
(Source: "Screaming Eagles Attack Village," by SFC Peter Varisano, U.S. Army Ctr. for Mil. Hist., from url: http://www.history.army.mil/art/hartgers_and_varisano/101st.jpg)

in the absence of direct supervision, can he as a rusher take full
advantage of a diversion, know when to get down, or move along
a recently discovered depression in the ground to pop up in a different location. Whenever he reappears in the same place he was

last seen, he has considerably less time to make his next move. If a defender already has his rifle aimed at this spot, how long will it take him to pull the trigger? Though doing such a maneuver correctly takes practice, it will quickly capture the momentum in any chance encounter. Chesty would have loved it.

If the squad leader were to try to orchestrate such a complicated multi-lane evolution, he would not only telegraph every rush, but ignore each rusher's limitations. Some will get up slower than their buddies. Others can't run as fast. That's why each must be allowed to do his own thing. Only when loosely controlled will he and his lane partner do as well. Then, as the next chapter will show, this lowly "buddy team" becomes the key to more powerful American units.

16 Personal-Decison-
___ Making Practice

● How does a junior grunt know what to do when all alone?

● Shouldn't he be given more advice on common scenarios?

Soon, each must shoot around his forward-moving buddy.

The Trusting Teenager

Junior U.S. infantrymen have always had the most sophisticated equipment, and training on how to use that equipment, in the world. New communication gear now makes them more rapidly aware of all commander decisions. Thus, the average American "snuffy" thinks himself ready for combat. As a recent high school graduate, he trusts his national security agency to have adequately prepared him. The below stories are purely fictional. They have been added to show that, despite all the organizational advances, this U.S. rifleman must remain fully proactive with his environment.

No More Sunrises

Darkness has now descended over the rain forest like an undertaker's cloak. Not even the stars are visible. As the mist rises from the jungle floor, the croaking of frogs, buzzing of insects, and calling of animals all blend into a dull roar. It's so dark that the U.S. perimeter guard sees only vague shapes within his assigned sector. Though he has pals less than 20 yards away on either side, he still feels isolated and uncomfortable. There are so many shadows, plants, and ground irregularities between them and him, that those 20 yards might as well be 2,000. His squad leader has the only Night Vision Goggles (NVGs), and his even more distant platoon leader the only thermal-imaging device. Under such conditions, neither piece of technology would do him much good anyway. The former needs ambient light, and latter can't see through bushes.

Raised in the city and with only six months in service, this American watchstander feels out of place in the woods. After being repeatedly told he is the best in the world, he still has his doubts. His uncle and grandfather have both talked of German, Japanese, North Korean, Chinese, and North Vietnamese infiltrators being able to close with a wide-awake sentry. The young American is tired. For weeks on end, he has been patrolling all day and manning a perimeter hole most of every night. Though only 130 pounds soaking wet, he has been lugging 100 pounds of mostly ammunition through every bog and thicket in the area. At first, he tries to analyze every sound and shadow. Then, his mind wanders back to Trish and home. For hours, he sits erect in his foxhole, moving in and out of "the here and now." Then it happens. He has failed to notice that the 12 bushes to his front have turned to 13. There's a "whoosh," a dull "thud," indescribable pain, a suspicion of betrayal, a gasping for breath, and then nothing at all. Pvt. Robert B. "Squirt" Ryan, Jr.—the pride of Bedford Falls—is gone. He will not have that family of which he and his high-school sweetheart had dreamed. He will not spend his Saturdays fishing with his best friend Bill. He will not become that fireman who would save all those lives.

Just Another "Non-Contact" Patrol

A different squad from Pvt. Ryan's company has been going through all the standard motions of security patrolling for weeks.

Their routes have been officer selected, and checkpoints all reached. Yet, they have made no contact in an area that battalion intelligence claims to be awash with enemy soldiers.

PFC Paul Smuckatella has been a member of this squad for quite some time. Here's how he describes its patrolling practices to a friend from a different platoon. Most legs are walked single file, with everyone doing exactly as the guy ahead has done. At any break (however small) in the brush, the whole squad stops as a couple of guys take their sweet time checking out the other side. Then, the patrol moves on in the same slow and methodical fashion. What usually results is a "stop-and-go" routine so mind numbing as to suck all motivation (to include any desire for contact) from all who participate. Their only remaining objective is to get back to base.

Of course, PFC Smuckatella and his buddies have rehearsed "immediate-action" drills for what to do when ambushed. Still, he wonders how that perfectly aligned (and fully upright) formation would fare against automatic weapons just over 50 yards away across ground that is perfectly level and thick with tall, spindly stemmed plants.

Paul's squad leader is a "lifer" who expects every fire team leader to closely follow both his orders and established procedure. That leaves little room for rifleman input. Still, Paul has a hunch that much will depend on his degree of involvement. While walking "point" one day (as the most recent "new guy"), he sees some things that don't add up: (1) cloudiness at a place in the stream that should have been clear; (2) strange indentations in the ground as from the ball of somebody's foot; and (3) grass bent over at regular intervals along one side of the trail. What wild animal or other natural phenomenon could have caused them? And what of that faint whiff of gun oil?

Nothing happened that day, but two weeks later all 13 members of a sister patrol were killed by enemy ambushers in the same general vicinity. Paul asked his squad leader about this. All he got for an answer was, "They weren't alert enough." PFC Smuckatella wasn't stupid. He suspected a low-ranking member of that patrol had also seen things he didn't understand, and then been reluctant to bother his "all-knowing" superiors with them. What that other E-2 had lacked in knowledge and assertiveness had helped to get him killed.

Decision Making Easily Added to Technique Drills

In top-down bureaucracies, the lowest ranks are generally thought to have the most to learn. That's why most U.S. combat veterans think their actions so insignificant to the war's eventual outcome. Though supremely loyal to "superiors," such a perception is far too modest. Whether a conflict is won against a "bottom-up" (Asian) adversary has mostly to do with the nonrates' contribution. For it is they who first encountered one of the foe's favorite strategies—"death by a thousand razor cuts." If that conflict then turned out badly, it was likely due to their being too little trained or trusted.

Though highly intelligent and inquisitive, the 18-year-old American recruit is no mind reader. If he is to perform "morally" in combat, he must be shown how most ethically to handle recurring situations. That can't be done *en masse* in some stuffy classroom. Each must be allowed to react to real-world challenges in mock combat. Below are some of the most productive movement technique drills for riflemen. To each has been added participant decisions—some tactical and others moral.

The Choice-Oriented Gurkha Trail

Marines have occasionally walked "Gurkha Trails" since the mid-1980's. Normally situated along a deep ditch to provide a ready-made backstop for short-range marksmanship, such a trail also tested the student's eyesight. Riflemen (closely followed by an instructor) moved up the trail one at a time. Each then tried to "double-tap" every upper-torso silhouette he encountered. (See Figure 16.1.) After all such targets were compared to the total, he was ranked against his peers and given a chance to improve his score.

Such Gurkha Trails would be just as good at determining point man potential, as training selectees. Along them can be positioned all manner of evidence. The visual might be a cigarette butt, audible the recording of an AK-47 coming off safe, and olfactory the smell of garlic. (See Figure 16.2.) To be really useful, such an exercise should also include footprints and other background-altering signs of human passage.

Figure 16.1: Instructor-Trailed Student Spots Silhouette
(Source: FM 7-8 [Aug. 84], p. B-2)

Figure 16.2: Enemy "Sign" Can Be Added to Any Gurkha Trail
(Source: FM 21-75 [1967], p. 8)

Figure 16.3: "Blanks Only" Trail for Real Encounters
(Source: "A Concise History of the Unites States Marine Corps 1775-1969," by Capt. William D. Parker, sketch by Capt. Donald L. Dickson, Hist. Div., HQMC,1970, p. 62)

Then, a separate "blanks only" trail could test/develop the security provider's capacity for moral-decision making. (See Figure 16.3.) After being told the grade depended on how quickly blanks could be fired into an appropriate target, he would meet various people coming along the trail from the opposite direction. The second of these people might be unarmed, and fifth trying to raise his hands. Shooting either by accident would earn this rifleman 10 demerits. (See Figures 16.4 and 16.5.) U.S. police regularly do this kind of training on live-fire ranges, but they more easily afford noncombatant "pop-up" targets.

Night Contact Training on "Blanks-Only" Trail

Among the point man's biggest challenges is how to deal with the lead scout of an approaching patrol. That's when the headquarters' tally of all friendly activity in an area becomes so important. The

Figure 16.4: Military-Aged Male with No Weapon
(Source: FMFM 2-1 [Mar. 1967], p. 115)

Figure 16.5: Enemy Combatant Trying to Raise His Hands
(Source: FM 7-8 [Aug. 1984], pp. F-1, I-9)

American point man has no time for allegiance determinations. Properly to function, he must assume the shadowy figure to be hostile and then follow a few close-encounter axioms.

If both men are fully exposed at night, he who shoots first will win. If the U.S. point man is behind cover or laying flat on the ground, there's a good chance the enemy soldier hasn't seen him. Then, the GI has other options: (1) signal back to the rest of his squad to set up a hasty ambush; (2) attempt to capture the foe; or (3) do nothing other than reporting the sighting. During a "non-contact" reconnaissance or objective approach, the latter may be unavoidable. (See Figures 16.6 and 16.7.)

Figure 16.6: Upright Point Man in the Open Has Little Choice
(Source: FM 7-8 [Aug. 1984], p. 5-23)

Figure 16.7: Prone Point Man with Cover Has Other Options
(Sources: FM 90-3 [Aug. 1977], p. 4-9; MCWP 3-35.3 [Apr. 1998], p. A-1)

Close Contact Easily Turns into a Grenade Exchange

When throwing live grenades in training, one stands behind a barricade. That much cover is seldom available in combat. Yet, the drill is still useful. Some people lack the physical coordination to hold on to a grenade long enough to toss it a few feet. The sooner they are identified, the better. Others forget to remove the safety clip before pulling the pin. Still, grenade fighting involves much more than that.

In actual battle, most grenade work looks like that in Figure 16.8. This prone position should first be practiced on unobstructed terrain against 55-gallon drums or large trash cans. A rounded object can almost never be lofted directly into an opening. It must be rolled in. Thus, to count as a hit, all practice tosses must only end in a "thunk." (See Figure 16.9.) From this prone position, a sideways-looping trajectory often results. In wooded terrain, its precise path would be impossible to predict. There, kneeling behind a stump or mound would be preferable. (See Figure 16.10.)

215

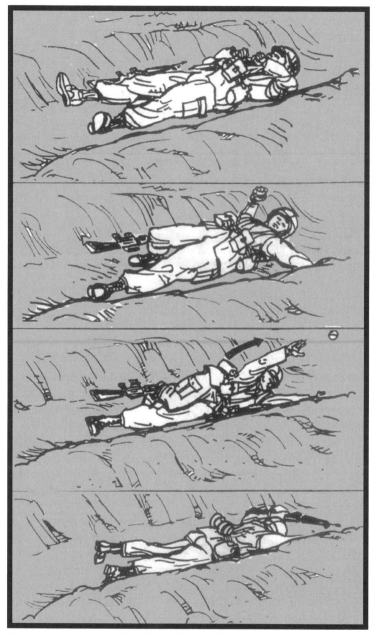

Figure 16.8: Prone Position Normally Only Way to Avoid Fire
(Source: FM 23-30 [Dec. 1988], p. 2-8)

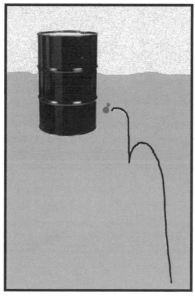

Figure 16.9: Easiest Way to Count Hits in Training
(Source: FM 23-30 [Dec. 1988], p. I-30)

Figure 16.10: Upright Position Provides More Accuracy in Woods
(Source: FM 7-8 [Aug. 1984], p. B-3)

Any soldier who has—more than once—bounced a practice grenade off a nearby tree (or had one slip out of his hand), should not be carrying the live variety in combat.

Fire and Movement More Useful Than Previously Thought

Since the mid-1980's, the leapfrog maneuver known as "fire and movement" has been considered ineffective against a deliberate defense. (See Figure 16.11.) Though its odds against barbed wire are not good, it is now known to work well with a barbed-wire-free strongpoint matrix. Because the front-row bunkers are protected by fire from those between and just to the rear, all bunkers in both rows must be invested at once. Otherwise, even Stormtrooper technique will be futile. That simultaneous investment most easily happens with a loosely controlled line of fire teams, each having its own parallel lane. Without any outside coordination, every other team would then suppress the indented bunkers, while those between crawl up on the forward emplacements. At no time is there any need (or risk to friendlies) of standoff supporting fires. This safer way of seizing a defense in depth was discovered by 4th Raider Battalion at Bairoko Harbor on New Georgia while without artillery and after a skipped aircraft mission. It was then intentionally repeated (no tanks, planes, or artillery requested) by 2/4—this same Raider unit's redesignation—while capturing the most difficult part of the Sugar Loaf Complex on Okinawa.[1]

When properly done, "fire and movement" can be extremely powerful. Still, it's more complicated than it looks. Bad habits result from a hastily designed drill. Rushers must not be required to move straight forward to give their partners an easier shot. That only makes the upright personnel better targets. The two-man lanes must be wide enough to permit a little zigzagging (like the Finns used to do).[2] This type of motion comes fairly naturally, as pieces of cover aren't often aligned. With enough practice, base-of-fire personnel have no more trouble shooting around partners veering off to the left or right. After all, the enemy may be traveling sideways as well.

Of utmost importance, each drill participant must be allowed to decide when to rush. Otherwise, the whole evolution becomes too predictable. Normally, he does so after a diversion elsewhere on the line. (See Figure 16.12.) Then, he must continually choose

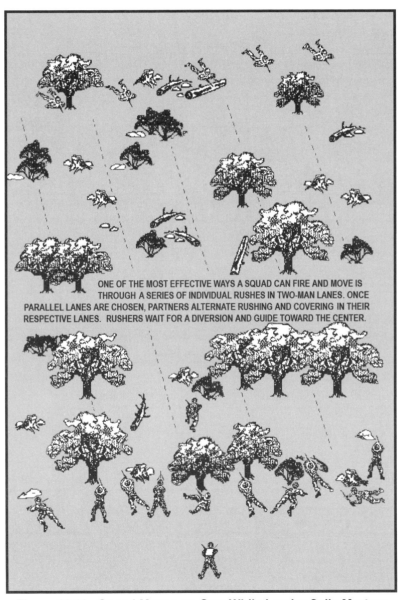

ONE OF THE MOST EFFECTIVE WAYS A SQUAD CAN FIRE AND MOVE IS THROUGH A SERIES OF INDIVIDUAL RUSHES IN TWO-MAN LANES. ONCE PARALLEL LANES ARE CHOSEN, PARTNERS ALTERNATE RUSHING AND COVERING IN THEIR RESPECTIVE LANES. RUSHERS WAIT FOR A DIVERSION AND GUIDE TOWARD THE CENTER.

Figure 16.11: Squad Moves on Own While Leader Calls Mortars
(Sources: MCI 03.66a [1986], p. 2-8; FM 22-100 [1983], p. 66; FM 7-70 [1986], p. 4-20; FM 7-11 [1978], p. 2-III-E-8.2; FMFM 6-7 [1989], p. 1-13; FM 5-103 [1985], p. 4-6)

Figure 16.12: Whole Trick Is Not Making Too Good a Target
(Sources: FM 22-100 [1983], p. 84; FM 7-70 [1986], p. D-24)

Figure 16.13: Move Elsewhere to Cover after "Hitting the Deck"
(Source: FM 21-75 [1967], pp. 26, 77)

Figure 16.14: "Up, He See's Me, I'm Dead"
(Source: FM 100-5 [1994], p. 56)

how long to remain upright. Normally, that's the time it takes his
nemesis to draw a new bead with his rifle—three seconds. The ver-
bal equivalent would be, "Up he sees me, I'm down." But, beware;
three seconds may not always be available.

Rushers Must Shift Location Every Time They Go Down

There are ways other than a face-first slide to get down after
each rush. Each participant must develop his own. One might
be a squatting skid to soften the forward flop. Then, to keep from
getting immediately shot when next appearing, each man must be
trained to crawl a short distance to some stump, rock, or depression. (See Figure 16.13.) Those pieces of cover will seldom be three
seconds apart, so trying to reach the next while still upright will be
a constant temptation. This technique-included deception should
force all defenders to draw a new bead every time a rusher pops up.
That means another three seconds with which to work. Some people
have instead learned to roll upon hitting the ground. That could
disorient them. Should any forget to change location altogether,
their time-keeping sentence would change. (See Figure 16.14.)

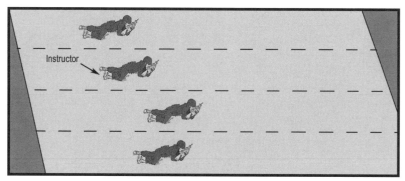

Figure 16.15: Crawl Racing
(Source: FM 100-5 [1994], p. 56)

Crawling Now Too Little Appreciated

Just as double-taps and grenade throwing are subsets of the "shooting" basic, so are short rushes and crawling important parts of the "moving" basic. Crawling has many uses in combat—from safely crossing a danger area to short-range infiltration. Fully to utilize it, every GI will need more technique and endurance. He should regularly take part in crawl races as part of his physical conditioning. This won't be viewed as harassment as long as the instructor participates. (See Figure 16.15.)

New Abilities Combine into Powerful Squad Technique

To see how "good" fire and movement can be practiced without participants ever having to return to the starting line, look at Figure 16.16. Then, Figure 16.17 shows how the final portion of this maneuver might look in actual combat. For advanced small-unit tactics, light infantrymen depend on proficiency in the subsets of what heavy infantrymen consider to be "bottom-line" (indivisible) basics.

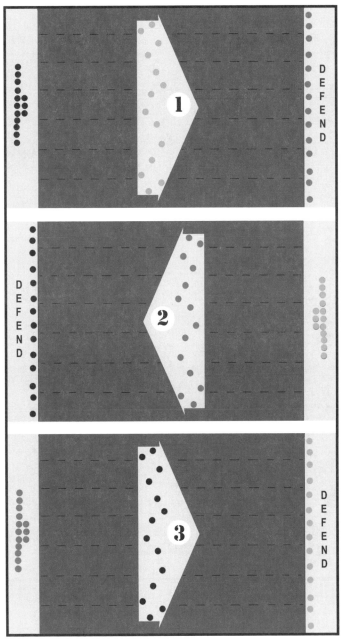

Figure 16.16: Back and Forth Squad Movement

Figure 16.17: Grenades Often Complete the Fire and Movement
(Sources: MCI 03.66a [86], pp. 2-8, 2-9; FM 22-100 [83], pp. 66, 84; FM 7-70 [86], p. 4-20; FM 7-11 [78], p. 2-III-E-8.2; FMFM 6-7 [89], p. 1-13; FM 5-103 [85], pp. 4-6,4-38; FM 7-11B3 [76], p. 2-VII-C-4.4)

"Battle Drills" for What Is Thought to Be Correct

All battle drills should determine how much each participant has learned from the drill. That's why they have the following format:

> Attention gainer and lecture.
> Demonstration and practical application.
>> Outdoors.
>> Blackboard or overhead projector.
>> Sand table with miniatures.
> Practical application testing.

Instead of the instructor being expected to make all students proficient, why not shift some of that responsibility to each student himself. More than one chance at practical application makes this possible. On succeeding tries, won't every individual work at improving their own performance?

As long as each student's ability is being measured, why not use their collective score to reassess technique effectiveness? No technique is perfect. The more it is practiced, the more refinements will become apparent. They may involve an overall design change to lower the risk, or just more execution leeway for each participant. The tactical proficiency of every maneuver will still be inversely proportional to the number of personnel it expends. If there is no way to determine student proficiency through simulated casualty assessment, then speed, stealth, deception, or some other measure of surprise are all available. With more surprise of the enemy comes less harm to friendlies.

One Exercise That Involves a Lot of Decision Making

Just as more than one pugil-stick-wielding adversary helps to develop the proactive fighter in Chapter 12, so can dealing with more than one way of getting shot. The best place for such an exercise is the city. Here, there are so many threats that one's individual movements can be as impossible to cover by fire, as they are to conceal. (See Figure 16.18.)

The urban attacker can be dinged from almost any direction and runs a much greater risk of movement signature, reflection,

225

Figure 16.18: Modern Wars Involve More Street Combat
(Source: "Street Fight," by SFC Elzie Golden, U.S. Army Center of Military History, from this url: http://www.history.army.mil/art/Golden/Image-11.jpg)

or unnatural shadow. With no way to camouflage himself, he is drawn inexorably toward dark indentations in the walls. There, the narrowest of cracks can still give him away. Without extensive practice on how to move in a built-up area, the young warrior has not long to live.

Neither can the urban attacker ever be fully covered by fire. While moving along a street, he can be shot from front, back, above, or below. He can be killed from 100 obvious places and any number of hidden ones. There is no muzzle flash from behind a curtain. If an enemy soldier were to shoot from atop a table inside an upper-story room, his sector of fire at street level could be so small as virtually to preclude return fire. Thus, while operating around the base of buildings, the urban attacker must defeat the first bullet himself—through how he moves.

When he goes indoors, his problems only magnify. There, he becomes more canalized and susceptible to ambush. He can be shot through walls, floors, ceilings, or any window from hundreds of yards away. As his pals can only keep an eye on hallways and stairwells, he must again stake his life on his own ability to move. Unfortunately, the official manuals are of little help in this regard. Only through field experimentation can the best ways of changing location be determined, and they will often vary between individuals.

So, preparing a rifleman for urban combat has largely to do with how much of a chance he has to train himself. Moving with any measure of safety takes continually reassessing complex terrain for multiple threats. For the best result, the instructor/facilitator concentrates on one or two threats at a time and then in combination. Every student will have unique limitations. By giving him back-to-back opportunities to improve against simulated fire, the instructor generates enough feedback to allow each individual to refine his own technique. Yet, there is an even bigger payoff from such training. Whatever the majority of students end up doing to deal with so complex a threat could usefully become the standard guideline (as opposed to required procedure).

Urban Drill Specifics

The students will be responsible for their own progress, so they must first be reminded of all the ways to get shot in the street or between buildings. Below are the most important:

(1) By bullets deflected along walls
(2) By long-range machinegun/sniper fire from up the street
(3) From upper stories or sewer grates to the front and rear
(4) From spaces between the buildings
(5) From windows, doors, or tiny wall openings (like air vents)
(6) Through closed doors or wooden walls

A pair of instructors find a deserted city street like one in the war zone. They issue rubber rifles. Then, each takes half of the student body and two augmentation aggressors to one end of this street. (See Figure 16.19.) All further actions will on opposite sides

227

Figure 16.19: Waiting Their Turn at Street Movement Practice
(Source: Courtesy of Sorman Information and Media, illustration by Wolfgang Bartsch, from "SoldF: Soldaten i falt," © 2001 Forsvarsmakten, Stockholm, pp. 22, 23)

and completely separate (to double the output). From each column, one student at a time moves up the street. He and his pals will get successive tries against three pairs of threats.

Each sends his aggressors to the other end of the street. One sits in the middle to simulate long-range machinegun or sniper fire. The other stands next to the store fronts to simulate wall-following bullets. Both will count three-second sight pictures of the moving student. (See Figure 16.20.) Then, that student is told how many times he died and sent back—behind the buildings—to the end of his column. After everyone has had two chances, they must face a less obvious challenge.

This time, the aggressors obscurely provide examples of the third and fourth threats. The instructor tells his students to avoid fire from between buildings and above or below street level. Hits for each are relayed by cell phone to the instructor. After learning of the scores, all try to improve.

Figure 16.20: Close Attention to All Openings in the Walls

(Source: Courtesy of Sorman Information and Media, illustration by Wolfgang Bartsch, from "SoldF: Soldaten i fält," © 2001 Forsvarsmakten, Stockholm, p. 289)

Figure 16.21: The Underappreciated Collective Opinion
(Source: FM 22-100 [1983], p. 164)

Finally, the aggressors provide hidden examples of the fifth and sixth threats. The instructor tells runners to elude fire from apertures and thin surfaces. All get two chances.

One or two days later, the same instructors deploy six augment aggressors apiece and work on various combinations of four threats, and then all six at once. Only one man can run the gauntlet every five minutes, so each column should be no longer than eight people. For more students, the command should set up identical courses at other locations. Similar packages can be designed for interior movement.

"Situation Stations" for What Is Not Certain

Sometimes, the government manuals give no clue as to how to handle a certain set of circumstances. Then, the instructor may be at a loss on to how to prepare his people. He needn't be. With an overall knowledge of infantry axioms, he has only to "facilitate" their collective opinion. (See Figure 16.21.) For example, he can discuss each sector's possibilities while walking a prospective perimeter with a group of 15 to 20. Whenever a student comment is consistent with the axioms and acceptable to the group, he adds it to the working solution. Nonrates are the first to die, so their group insights are of great tactical value. After interacting in this way with the first few "sticks" of personnel, the instructor would become fairly confident of the answer. He who routinely "mines" the nonrates' solution to short-range scenarios will eventually become a light-infantry expert. (See Figure 16.22.)

For the True Professional

This chapter has been of most interest to recent U.S. recruits (prospective casualties) or entry-level instructors. Most line-unit leaders have only skimmed over its details. They have too often been told of their personal abilities being the deciding factor in combat. Above the instructor hall exit at Quantico's Basic School is a quote from a mid-19th-Century Naval Commodore to the effect that officers win wars.[3] In a way, they do. But, with infantry units, it is not as most people think. A confining boat on the open ocean is quite different from a rifle company with room to spread out in heavily foliated terrain. How closely a naval captain controls his crew more directly influences how well they will do in an enemy contact. The infantry unit's opposition is likely to be well hidden, decentralized, highly opportunistic, and deception oriented. Only through bottom-echelon skill and initiative can such an adversary be consistently beaten. This takes looser control over the friendly contingent.

In essence, infantry leaders mostly win wars through how they prepare their people ahead of time. Where everyone still has room to maneuver, the force with the most skilled and assertive tiny elements will generally win. Such elements require seasoned light infantrymen whose training has already instilled initiative, tacti-

Figure 16.22: Best Instructors Learn from Their Students
(Source: "STRAC Instructor," by SFC Elzie Golden, U.S. Army Center of Military History, from this url: http://www.history.army.mil/art/Golden/Image-5.jpg)

cal-decision making, and upwards dissemination of intelligence. Within U.S. units, the prerequisite level of tactical expertise will not be possible until junior-enlisted members help to design their own small-unit maneuvers.

17 Troops Must Help to _ Design Own Moves

● Why must riflemen contribute to squad maneuvers?

● How are their ideas most easily collected?

No Private in his right mind would practice being this fully exposed.

What It Will Take to Improve Small-Unit Maneuvers

By November 1943, U.S. Marine units had all come to depend on the same standardized method for assaulting an enemy pill-box. It has been variously described as "Blind 'em, Burn 'em, and Blast 'em." Small-arms fire supposedly drove bunker occupants from their firing ports long enough for napalm-carrying Marines to move up and direct flame through the openings. Then, demolition men approached to toss in satchel charges.[1] Unfortunately (as indicated by the Tarawa reenactment in "Sands of Iwo Jima"),[2] the first part of this technique was overly optimistic. (See Map 17.1.)

During the initial suppression phase, bunker occupants often did too much shooting to allow for the last two steps. In fact, a former scout sniper—William Deane Hawkins—became so incensed with the technique's shortcomings that he went on "a one-man rampage [on Tarawa], attacking pillbox after pillbox" by himself. He did so by "crawling up to them in the sand, firing into the gunports, and [then] tossing grenades inside." (See Figure 17.1.) He was finally killed by mortar fire and later awarded the MoH.[3]

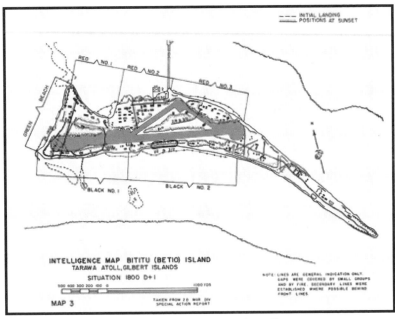

Map 17.1: The Tarawa Landing
(Source: Wikipedia Encyclopedia, s.v. "The Battle of Tarawa," with image designator "File_USMC-M-Tarawa-3.jpg")

Figure 17.1: Firing Ports on a Typical Japanese Bunker
(Source: TM-E 30-480 [1944], p. 159)

Part of the problem was that the Marines of this era had no easily carried automatic weapon with a high enough rate of fire to completely fill a bunker aperture with bullets. While preparing for the Iwo Jima landing, a former armorer by the name of Cpl. Tony Stein decided to personally correct this oversight. Upon hitting the beach at Iwo, he successfully assaulted several bunkers with his tailor-made "Stinger" before the odds finally caught up with him at Hill 362A. (Refer back to Maps 13.1 and 13.3.)

For conspicuous gallantry and intrepidity at the risk of his life above and beyond the call of duty while serving with Company A, 1st Battalion, 28th Marines, 5th Marine Division, in action against enemy Japanese forces on Iwo Jima, in the Volcano Islands, 19 February 1945. The first man of his unit to be on station after hitting the beach in the initial assault, Cpl. Stein, armed with a personally improvised aircraft-type weapon, provided rapid covering fire as the remainder of his platoon attempted to move into position. When his comrades were stalled by a concentrated machinegun and mortar barrage, he gallantly stood upright and exposed himself to the enemy's view, thereby drawing the hostile fire to his own person and enabling him to observe the location of the furiously blazing hostile guns. Determined to neutralize the strategically placed weapons, he boldly charged the enemy pillboxes 1 by 1 and succeeded in killing 20 of the enemy during the furious single-handed assault. Cool and courageous under the merciless hail of exploding shells and bullets which fell on all sides, he continued to deliver the fire of his skillfully improvised weapon at a tremendous rate of speed which rapidly exhausted his ammunition. Undaunted, he removed his helmet and shoes to expedite his movements and ran back to the beach for additional ammunition, making a total of 8 trips under intense fire and carrying or assisting a wounded man back each time. Despite the unrelenting savagery and confusion of battle, he rendered prompt assistance to his platoon whenever the unit was in position, directing the fire of a half-track against a stubborn pillbox until he had effected the ultimate destruction of the Japanese fortification. Later in the day, although his weapon was twice shot from his hands, he personally covered the withdrawal of his platoon to the company position. Stouthearted and indomitable, Cpl. Stein, by his aggressive initiative sound judgment, and unwavering devotion to duty in the face of terrific odds, contributed materially to the fulfillment of his mission, and his outstanding valor throughout the bitter hours of conflict sustains and enhances the highest traditions of the U.S. Naval Service.[4]

— MoH Citation for Tony Stein

Why These Two Marines Took Such a Risk

In effect, a pair of low-ranking personnel had just tried to show their massive organization how better to dismantle the building block of every Japanese defense. Tony had been a longtime Paramarine (an outfit formed about the same time as Carlson's Raiders) and thus afforded more initiative than the average infantry NCO. As a former scout sniper, Hawkins had enjoyed more autonomy as well. However, it's doubtful that they were the only Leathernecks to have realized the assault procedure's limitations. By a simple show of hands, any group of riflemen could have shared this unfortunate truth with their commander.

Hawkins' and Stein's courageous "demonstrations" would do more than just add firepower to the method. Their input was to help change its entire format. In future years, Marines wanting to take down a machinegun bunker would have rapid-firing Squad Automatic Weapons (SAWs), highly accurate grenade throwers, and more powerful bazooka-type ordnance. Gone were the days of someone trying to approach one with a napalm bomb on his back. Yet, Hawkins' hard-won lesson—that there was microterrain to the front of many fortified emplacements that allowed a crawling approach—would go largely unheeded into the modern era. Why is the question.

Not only do all those little ripples in the ground protect assault troops from objective bunker fire, but often from that of its matrix partners. In a defense belt, the front of every pillbox will be crisscrossed by grazing fire from those on both sides behind it. That makes any kind of upright attack very dangerous. 4th Raider Battalion had done a lot of crawling at Bairoko, and its redesignation—2/4—again at Okinawa's Sugar Loaf Complex.[5] In Vietnam, 2/4 and 3/4 would both opt for the prone-fire-team approach over the risky sledgehammer.[6] Yet, most U.S. battalions have stuck with the upright, firepower-spewing attack style. They may have continued to ignore the ground irregularities for a reason. Ordnance-rich commanders would principally seek fire superiority. One wonders what a little troop input now might do for their track record.

Let The Troops Do What?

The newest members of a top-down organization are almost never asked for their viewpoint. Still, it is they who will pay the ultimate price for any ill-advised move on an active battlefield.

237

When pressed, many would admit to the extreme danger of certain maneuvers in their manuals. Advancing upright into enemy machinegun fire is only one of them.

Throughout the Marine Corps' long and illustrious history, the rank and file have seldom been allowed any input on their own tactical techniques. From 1997 to 2006, there was a "bottom-up" supplement to squad training that was tested in 41 battalions throughout the Fleet. It involved a "Tactical Demonstration" by the lowest ranks of each company.[7] Within that Demonstration, every rifleman (and his pals) got the chance to show peers and superiors

Figure 17.2: Marines Had to Learn Fast on Guadalcanal

alike how better to conduct small-unit maneuvers.[8] Yet, nowhere in the HQMC-dictated curriculum has anything commensurate since been authorized. To find an officially sanctioned example, one must go all the way back to 1942. (See Figure 17.2.) Even Evans Carlson's all-hands "*Gung Ho* Sessions" on Guadalcanal only occasionally dealt with tactical issues. The Sessions were mostly for admitting mistakes, explaining plans, and building initiative.[9] Yet, Carlson was also attempting to learn from his most junior subordinates. If a fire team was to attack a pillbox, he wanted ideas on how best to take that pillbox.[10] Unfortunately, there were few bunkers of any type on Guadalcanal, and thus not much need for advanced assault technique. Only once during these "*Gung Ho* Sessions" does the record show a Raider disagreeing with his company commander over a just-practiced maneuver.[11] Nor is there any evidence of Carlson getting a show of hands on someone's tactics proposal, or of that proposal subsequently being tested in mock combat.[12]

> Borrowing an idea from China, Carlson frequently has what he calls "kung-hou" meetings. . . . Problems are threshed out and orders explained.[13]
> — *New York Times,* 1942

> During World War II, Lt.Col. Evans F. Carlson's 2nd Raider Battalion held regular "*Gung Ho* meetings" in which all Marines and sailors had an equal voice in working out issues.[14]
> — *Marine Corps Gazette,* 2003

Gung Ho Session May Have Briefly Reappeared in Vietnam

During the Vietnam War, one Lima Company member of 3/4—a direct descendent of 3rd Raider Battalion—remembers participating in something like a *Gung Ho* Session.[15] But in 1/4, most "snuffy" input had slowly to wend its way up the enlisted ladder.[16] (See Figure 17.3.)

The NVA's "*Kiem Thao* Session" had also come from the Chinese model. In fact, it may have more closely followed it. A firsthand observer has described Mao's soldiers as "constantly studying their mistakes, and improvising methods to . . . offset modernized equip-

Figure 17.3: The Inflexible Steamroller
(Source: "The American Soldier, 1966," U.S. Army Center of Military History, from this url: http://www.history.army.mil/images/artphoto/pripos/amsoldier/5/1966.jpg)

ment."[17] That means some of the NVA's most effective strategies had come from nonrate suggestions—like how whole units could suddenly disappear into the ground.

A Much Needed Modification to Mao's Method

No matter the subject, allowing someone to voice their opinion in a group setting will help them to develop initiative. Still, with head-strong young Americans, any discussion of past mistakes should probably be limited to squad activity. Ideas on how better to land the landing force, conduct barracks life, or lead a platoon would only cause them to lose their focus. While a constantly "bitching" rifleman is a happy rifleman, his most valuable insights are of a tactical or ethical nature.

If this kind of group input were systematically gathered at the company level for wartime scenarios and then tested against simulated-casualty assessment, something quite special could emerge. That something would—for starters—be tactical parity with all Asian Communist, criminal, and other "bottom-up" adversaries. Throughout the world wars, both German and Japanese line infantry squads also had better movement techniques than their American counterparts.[18]

In a modern 4GW setting, small-unit proficiency becomes absolutely essential. The lack thereof in Vietnam and Afghanistan has already led to major embarrassments for America. If it is still missing from a low-intensity WWIII, the Pentagon's "big-picture thinkers" may become directly responsible for the unthinkable.

The brass is very good at taking power away [from lowest-echelon personnel]. It makes them feel like they are in charge.[19]

— *Criminal Minds,* ION TV, 9 December 2013

Combining Tactical Excellence with Moral Certitude!

Also possible from rifleman input is a continual check on the morality of the fighting. Then, through the Holy Spirit within, a more comprehensive "Ethical Indoctrination," and the morality-monitoring role, all American squad members would more easily

Figure 17.4: Christian Values Played Key Role in Nazis' Defeat
(Source: "Military Necessity," by A.Bohrod, U.S.Army Ctr.of Mil.Hist., posted April '98, from url: http://www.history.army.mil/images/artphoto/artphoto/artchives/1998/avop0498_1.jpg)

handle every challenge (4GW or otherwise). Though hastily trained and overcontrolled in Europe during WWII, their predecessors had still found a way to defeat the more tactically proficient Germans. That "way" must have been why a 1944 U.S.-Army-commissioned painting was entitled "Military Necessity." (See Figure 17.4.)

I am the way, the truth, and the life (sayeth the Lord).[20]

Afterword:
No Minor Oversight

The Overall Problem

As riflemen take the most risk in war, they are—collectively—one of the best sources of short-range tactical knowledge. In any Westernized unit, such a conclusion flies in the face of its rank and control structure. Many of the more legendary Marines—like Gen. Chesty Puller or Sgt. Toney Price—got that way by occasionally following only the "spirit" of their orders. Now, all U.S. Marine and Army leaders must also decide whether their job is to perpetuate Stateside bureaucracy or win overseas conflicts. The two are not the same; nor will they always mesh. Of late, America's military establishment seems more interested in total obedience than in delegating enough authority to become truly proficient at ground combat.

If this country is to win any more wars (including WWIII), all enlisted infantrymen (and their commissioned leaders) must follow through on the Posterity Press initiative. Its ways of more effectively training and operating are not particularly new or revolutionary, just a more detailed look at what the troops—as a whole—have always known. That's why the books sold so well during the war in Iraq. After the U.S. rank and file realized that the truth about their potential would little affect on their status, many have again lost heart.

The main difference between an Eastern and Western army is the greater extent to which the former trains and then utilizes its lowest-ranking infantrymen. What may now look like insignificant minutia to "big-picture" Pentagon thinkers is actually the way to correct this shortfall. For every "basic" U.S. infantry skill, Asian riflemen study numerous subsets. Besides moving upright, they learn how to crawl silently through dry leaves and swampy ground. Yet, only to the Iowa teenager about to assault an enemy machinegun, does such attention to detail hold much meaning.

Without the "collective pressure" of everyone forced—by organizational priorities—to discover this Pentagon deficiency the hard way, it may never be corrected. Whether "vets" or recent enlistees, all have every right to demand more comprehensive rifleman training.

The Depth of the Shortfall

Most of America's special operators still follow the U.S. Army's way of doing things—two fire teams per squad and simple "bounding-overwatch" maneuvers. To employ the deception that MW makes possible, they would minimally need a third fire team. Except for Sea, Land, and Air (SEAL) Team Six—which was "playing pickup basketball" during the Bin Laden seizure[1]—and some Army Rangers, few American commandos qualify as light-infantry experts. Their Asian Communist counterparts don't have such a handicap. That's why one of their parent units is called "The North Korean Light Infantry Training Guidance Bureau."

The fire team concept was copied from Mao by Carlson and then adopted by Edson. The new book—*Gung Ho!*—closely follows the Marine Raider battalions and their infantry redesignations through some of the heaviest fighting of WWII and Vietnam. Their highly refined fire team maneuvers would still be useful today (all that electronic paraphernalia doesn't make contemporary GIs any less visible). In fact, the Pentagon's new plan to use tiny teams as force multipliers throughout the world will not work without this kind of light-infantry expertise. Though Carlson's Raiders were doing a close facsimile of MW at the squad level on Guadalcanal, most of America's special operations and line infantry community still can't. 4th Raider Battalion (also Maoist in format) discovered on New Georgia how an advanced strongpoint matrix (bunker complex) can be more easily taken without any tanks, planes, or artillery. Their line infantry successor—2nd Battalion, 4th Marines (2/4)—proved that again at Okinawa's infamous Sugar Loaf Complex. Only necessary is a staggered row of fire teams "working together" from within parallel lanes.

Of late, many have assumed all U.S. wars over and Posterity Press books obsolete. Nothing could be further from the truth. Most of those conflicts have only shifted into lower-intensity or non-martial arenas. Hundreds of semi-autonomous infantry (or commando) squads will be needed for the Pentagon's worldwide containment strategy to succeed. Unlike Asian Communist special operators, "America's best" are still stuck with elementary and electronics-encumbered "approach," "assault," and "withdrawal" techniques. Anyone who still doubts this has only to Google the term *"dac cong"*—the Vietnamese word for commando or sapper. To make matters worse, most U.S. special operators still have no way to hide when their capture seems imminent (the E&E part of UW).

Through the Pentagon's "top-down" approach to everything, a chronic deficiency has developed at one of its bottom echelons, with regard to squad, fire-team, buddy-team, and rifleman tactics. The "state of the art" on such things (also known at light-infantry technique) now comes mostly from Asia, with a few Western memories. That's why Posterity Press titles will always be relevant. They contain the latest light-infantry research. As again proven by the Afghan experience, the wave of the future is in "distributed light-infantry operations," not tighter overall control or higher-tech ordnance. (See Figure A1.1.)

Figure A1.1: Things Go Better When All GI Talent Fully Utilized
(Source: "Two Soldiers on Night Patrol," by Harold Von Schmidt, U.S. Army Center of Military History, from this url: http://www.history.army.mil/art/Posters/WWII/1-38-49.jpg)

Today's warriors don't easily admit to such a deficiency. Having worked hard and accomplished much, they honestly believe that the advances in small-unit communications alone have allowed them to do things their predecessors couldn't. For them, two comments are offered: (1) this nation's profound thanks for a job well done; and (2) the following reminder.

There is only one way for a squad-sized patrol to consistently do well during chance contact. As in football, each of its members must be allowed to slightly deviate from what the group has practiced for a similar scenario. No amount of headquarters advice can keep pace with locally changing circumstances. Nor will instantaneous communications between squad members automatically produce a winning formula. It takes time to build a tactically effective maneuver. One attempted on the spur of the moment while trying to get the better of a thinking adversary would have only marginal odds of success. As the necessary coordination was in progress, the local situation would be changing. That's why all "brainstorming" must be done in advance, be based on past enemy performance, and then be slightly modified by squad members as applied in actual combat.

It is for these reasons that the 4th Raider and 2/4 way of assaulting a barbed-wire-free defense matrix is still more effective than a high-tech steamroller. Equipment gets damaged, and supporting-arms fire still comes too close. The WWII discovery only takes hand-and-arm signals between lanes and each man's personal weapon. That the fighters in adjacent lanes are coordinating among themselves without any overall control is what allows them to suppress the indented bunkers long enough for the forward emplacements to be taken. No amount of expert supervision could duplicate such a complicated feat.

The same holds true for a barbed-wire-protected strongpoint. But here, the best solution is even older—a prerehearsed WWI assault technique (or some variation) of German design.

So, here exists the central paradox that America's arms manufacturers and control advocates are in no hurry to admit. What happens at short range in combat is less affected by technology than other aspects of warfare. Better communications and target acquisition do not increase the individual soldier's chances all that much. Still just as visible as before, he still requires more comprehensive basics and state-of-the-art (however old) squad movement technique.

What It Would Take to Finally Fix This Problem

Every American Marine would love for his (or her) esteemed Corps to go on forever. Within the U.S. security establishment, that most easily happens where there is some unique capability. When "top-down" bureaucracies too greatly value the *status quo,* they often fail to make all the necessary changes. Despite a growing body of evidence that light infantrymen can hold their own—with acceptable casualties—in conventional battle, there are still no "truly light" infantry units anywhere within the U.S. Department of Defense (DoD). Even its commandos have trouble sneaking all the way up on a "woods-wise" quarry. If one of the service branches were now to develop a *bona fide* light-infantry capability, its future would be virtually guaranteed.

Though sometimes called "light," all U.S. infantry units are more accurately of the "heavy" or "line" variety, in which operating one's equipment and following orders take precedence over everything else. Light infantrymen don't need overwhelming firepower. They rely instead on small-unit initiative and tactical surprise. This alternative way of fighting should have been (but wasn't) of interest to the Army's airborne divisions. Neither has it been embraced (except by the short-lived Raiders) by the service branch that provides spur-of-the-moment amphibious landings. On the totally exposed battlefields of the future, tiny dispersed elements will more easily survive. Here, light-infantry expertise will be at a premium. Consider Figure A.1 again in light of the indirect (and mostly electronic) attempt—since November of 2011—to find Joseph Kony in Africa.

Instead of tightening control over every aspect of training to make things more orderly, Headquarters Marine Corps (HQMC) should try more experimentation with squad maneuvers at the company level. America's only "distributed ops" capability would do more for Corps' longevity.

The Standard Rebuttal

Most within America's special-operations and infantry communities believe the above-mentioned deficiency to have been already addressed. Their assessment of current U.S. capabilities goes something like this. "Sensors, robots, and drones will precede all future ground attacks. As a result of the new Global War on Terrorism (GWOT) mind-set, the associated assaults will be 'shaped' by intelligence actions: (1) psychological operations; (2) key leader engagements; (3) Signals Intelligence (SIGINT) exploitation; and (4) deception. [In effect], Marine Special Operations Battalion "door-kickers" won't need any improvements to their curriculum. They are already trained in initiative-based tactics, [combat] thinking, and [spur-of the-moment] decision making. Before any assault, they are fully apprised of all battlefield geometry—to include the microterrain they will be encountering. Surprise and stealth are part of their approach march. If it is lost, they will speed things up and counter with massive amounts of firepower. That makes the assault more lethal and concentrated than before."[2]

Apparently Acknowledged

While the claim to advance warning of every rock, bush, and ground undulation (presumably through overhead observation) seems a little too good to be true, the above assessment ("conventional wisdom") is not without merit. It does seem to concede that American forces will eventually have to disperse or fall victim to enemy target-acquisition technology. Once spotted in bivouac by an opposition drone, an entire American rifle company could be wiped out by makeshift rockets. This "conventional" viewpoint also makes provision for how those widely dispersed American elements could remain safe from piecemeal annihilation. They would all have their own (dedicated) air and artillery assets. Where scores of units got into trouble at once, that would be a sizable undertaking. Often in Vietnam, an entire company couldn't get air support when their attack objective became heavily contested.[3]

Where Are the Holes in Such Reasoning

What has been so hard for proud Americans to accept is that "bottom-up" Eastern societies more easily field semi-autonomous (or light) infantry squads. After disregarding frontline-war-veteran input and preferring headquarters conclusions, Western societies develop shortfalls the lowest echelons of their defense agencies.

No one in charge can remember how things used to get done, and the official manuals/chronicles contain too few details. As a result, everything within the U.S. military is always wonderful, all solutions underway, and the current generation of troops stronger than the last. All who have tried to make it a career can attest to this steady diet of optimism.

Though today's special-operation "door kickers" may get some initiative-based instruction, it's not as carefully thought out as that in Chapter 12. Whether they qualify as light-infantry experts is not difficult to ascertain. Ask one to describe in detail a WWI Stormtrooper assault and its Vietnam equivalent; or how to move unseen between enemy foxholes. If his first response makes no mention of concussion grenades, and the second of promising "gaps" in enemy lines, then that U.S. special operator is no expert of light-infantry assault.

The same type of test can just as easily determine the GI's level of defensive know-how. What role did each squad leader play in the Germans' "Elastic Defense in Depth" of 1917? [4] How did NVA troops keep—for four days—a fully supported U.S. battalion from crossing a narrow street in Hue City's Citadel? [5] If he makes no mention of squad leaders deciding when to abandon their positions, his light-infantry defense expertise is also missing.

During a heavily contested attack, no amount of firepower can take the place of a little surprise on the part of the attacker. All it takes to thwart a nearby row of assault troops is suddenly to thrust one's AK-47 up sideways out of a hole (with only hands showing) and sweep it across them on full automatic. Though the need for proficiency at this scale is nothing really new, it is almost never mentioned in after-action reports. That's because those reports have been so hastily composed in the heat of battle. On such occasions, commanders seldom take the time to dwell on what the enemy has done better.

Any "big-picture" assessment will always promise more situational detail that it can actually provide, and then fail to focus on what happens without it. That's what makes too much optimism such a hidden curse for "top-down" bureaucracies. For the lone U.S. gladiator, the most serious threat will continue to come from five-yards away. By better focusing on his preparation, U.S. infantry branches could partially override the above-mentioned side effect of their organizational structure.

Figure A1.2: Those Who Actually Won It
(Source: "The 28th Division in Paris," by Harry A. Davis, U.S. Army Center of Military History, from the following url: http://www.history.army.mil/art/A&I/28th.jpg)

The Ongoing Challenge

Being "the best in the world" at the small-unit and individual level takes constant effort, because of the direct involvement with an adversary (normally Asian) who must necessarily depend on this type of expertise. Here, too much U.S. pride can lead to important circumstances being ignored during all the tiny engagements that go into making any war. That may be why America hasn't decisively concluded one since WWII. (See Figure A1.2.)

For all U.S. service personnel working six days a week, always supporting the "team," and being made constantly afraid for their careers, this unfortunate reality is not easy to embrace. Yet, to properly safeguard their subordinates, they will have to. Because of the Communists' edge at short range, the future of the world may lie in the balance. Since the Vietnam War, it has become increasing clear that America's security establishment cannot defeat any "bottom-up" (criminal or Asian-oriented) foe without first allowing more initiative from its own lowest echelons.

Appendix:
Korean War Sighting

Figure A2.1: Michael the Archangel
(Source: Wikipedia Encyclopedia, s.v. "Michael (archangel)," with image designator "File: Michael4.jpg")

Michael, Michael of the morning,
Fresh chord of Heaven adorning,
Keep me safe today,
And in time of temptation
Drive the devil away. Amen.[1]

Below is a letter written by a young Marine to his mother after being wounded in 1950 on a Korean battlefield. After talking with the leader of his patrol, Navy Chaplain — Father Walter Muldy — has attested to the veracity of its contents.[2] (Before reading, look closely at Figure A2.1.)

Figure A2.2: This Marine Had Been Praying
(Source: "Strong Men Weep," by David Lax, U.S.Army Ctr.of Mil.Hist., posted Oct.'01, from url: http://www.history.army.mil/images/artphoto/artchives/2001/avop10-01_2.jpg)

Dear Mom:

I wouldn't dare write this letter to anyone but you because no one else would believe it. Maybe even you will find it hard but I have got to tell somebody.

First off, I am in a hospital. Now don't worry, ya here me, don't worry. I was wounded but I am okay—you understand? Okay. The doctor says that I will be up and around in a month.

But this is not what I want to tell you.

Remember when I joined the Marines last year: remember when I left, how you told me to say a prayer to Saint Michael every day. You really didn't have to tell me that. Ever since I can remember you always told me to pray to [ask for the intercession with the Redeemer of] Saint Michael the Archangel. You even named me after him. Well, I always have.

When I got to Korea, I prayed even harder. Remember the prayer that you taught me. "Michael, Michael of the morning, fresh cord of Heaven adorning." You know the rest of it. Well I said it every day. Sometimes when I was marching or sometimes resting. But always before I went to sleep. I even got some of the other fellas to say it.

Well, one day I was in the advance detail way up over the front lines. We were scouting for the Commies. I was plodding along in the bitter cold, my breath was like cigar smoke.

I thought I knew every guy in the patrol, when alongside of me comes another Marine I have never met before. He was bigger than any other Marine I'd ever seen. He must have been 6 foot 4" and built in proportion. It gave me a feeling of security to have such a body near.

Anyway, we were trudging along. The rest of the patrol spread out. Just to start a conversation, I said, "Cold, ain't it?" And then I laughed. Here I was with a good chance of getting killed any minute and I am talking about the weather.

My companion seemed to understand. I heard him laugh softly.

I looked at him [and said], "I have never seen you before. I thought I knew every man in the outfit."

"I just joined at the last minute," he replied. "The name is Michael."

"Is that so," I said surprised. "That is my name, too."

"I know," he said and then went on, "Michael, Michael of the morning . . ."

I was too amazed to say anything for a minute. How did he know my name, and a prayer that you had taught me? Then, I smiled at myself, every guy in the outfit knew about me. Hadn't I taught the prayer to anybody who would listen? Why now and then, they even referred to me as Saint Michael.

Neither of us spoke for a time and then he broke the silence. "We are going to have some trouble up ahead."

He must have been in fine physical shape for he was breathing so lightly I couldn't see his breath. Mine poured out in great clouds. There was no smile on his face now. Trouble ahead, I thought to myself, well the Commies are all around us, that is no great revelation.

Snow began to fall in great thick globs. In a brief moment the whole countryside was blotted out. And I was marching in a white fog of wet sticky particles. My companion disappeared.

"Michael," I shouted in sudden alarm.

I felt his hand on my arm, his voice was rich and strong, "This will stop shortly."

His prophecy proved to be correct. In a few minutes the snow stopped as abruptly as it had begun. The sun was a hard shining disc.

I looked back for the rest of the patrol; there was no one in sight. We [had] lost them in that heavy fall of snow. I looked ahead as we came over a little rise.

Mom, my heart stopped. There were seven of them. Seven Commies in their padded uniforms and jackets and their funny hats. Only there wasn't anything funny about them now. Seven rifles were aimed at us.

"Down Michael," I screamed and hit the frozen earth.

I heard those rifles fire almost as one. I heard the bullets. There was Michael still standing.

Mom, those guys couldn't have missed, not at that range. I expected to see him literally blown to bits. But there he stood, making no effort to fire himself. He was paralyzed with fear. It happens sometimes, Mom, even to the bravest. He was like a bird fascinated by a snake.

At least, that is what I thought then. I jumped up to pull him down and that was when I got mine. I felt a sudden flame in my chest. I often wondered what it felt like to be hit; now I know.

I remember feeling strong arms about me, arms that laid me ever so gently on a pillow of snow. I opened my eyes, for one last look. I was dying. Maybe I was even dead. I remember thinking, well, this is not so bad.

Maybe I was looking into the sun. Maybe I was in shock. But it seemed I saw Michael standing erect again, only this time his face was shining with a terrible splendor.

As I say, maybe it was the sun in my eyes, but he seemed to change as I watched him. He grew bigger, his arms stretched out wide, maybe it was the snow falling again, but here was brightness around him like the wings of an angel. In his hand was a sword—a sword that flashed with a million lights. [Refer back to Figure A.2.]

Well, that is the last thing I remember until the rest of the fellas came up and found me. I don't know how much time had passed. Now and then I had a moment's rest from the pain and fever. I remember telling them of the enemy just ahead.

"Where is Michael," I asked.

I saw them look at one another. "Where's who?" asked one.

"Michael, Michael, that big Marine I was walking with just before the snow squall hit us."

"Kid," said the sergeant, "You weren't walking with anyone. I had my eyes on you the whole time. You were getting too far out. I was just going to call you in when you disappeared in the snow."

He looked at me curiously [and said], "How did you do it, kid?"

"How'd I do what?" I asked, half angry despite my wound. "This Marine named Michael and I were just . . ."

"Son," said the sergeant kindly, "I picked this outfit myself and there just ain't another Michael in it. You are the only Mike in it."

He paused for a minute, "Just how did you do it, kid? We heard shots. There hasn't been a shot fired from your rifle. And there isn't a bit of lead in them seven bodies over the hill there."

I didn't say anything, what could I say? I could only look open-mouthed with amazement.

Figure A2.3: Just Ahead of Mike Were Seven Dead Reds
(Source: "The Sunshine Division in Korea," by Rick Reeves, poss. ©, from url: http://www.nationalguard.mil/resources/photo_gallery/heritage/images/sunshinedivision.jpg)

It was then the sergeant spoke again, "Kid," he said gently, "every one of those seven Commies was killed by a sword stroke."

That is all I can tell you, Mom. As I say, it may have been the sun in my eyes, it may have been the cold or the pain. But that is what happened.

Thanks for teaching me the prayer, Mom. I think it saved my life. I'll be home for Christmas.

Love,
Mike [3]

Notes

SOURCE NOTES

Illustrations:

Cover art is from U.S. AIR FORCE (www.af.mil). Entitled "Aim High," by Tech.Sgt. Cody Vance, this graphic was drawn by a paid member of the U.S. Armed Forces and is thus in the public domain. Its image designator, afg_030409_002.jpg, was retrieved in early 2006 from the following url: http://www.af.mil/shared/media/ggallery/hires/afg_030409_002.jpg.

Image on page vi from U.S. GOVERNMENT site and so in public domain.

Images from U.S. ARMY CENTER OF MILITARY HISTORY ARTWORK (http://www.history.army.mil/html/artphoto/artwork.html). Pages xxv, 37, 54, 78, 110, 242, 249, 254 from Archives jpg's: "0607-2," "Image-12," Hill_609," "1006-4," "0507-3," "avop04-98_1," "28th," "avop10-01_2." Pages xxvi, 96, 121, 187, 240, 245 from Topics jpg's: "4_229_46," "Rock," "Lend_the_Way," "avop06-00_2," "amsoldier/5/1966," "1-38-49." Pages 52, 60, 93, 113, 125, 205, 226, 231 from Staff jpg's: "Al inODA563," "Tracking," "61_27_45," "Supertroop," "Cav Trooper," "101st," "Image-11," "Image-5." See each caption for title, artist, and exact url.

Images on pages 3, 4, 13, 22, 25, 31, 43, 44, 49, 59, 66, 74, 77, 97, 101, 115, 190, 197 (both), 207, 233, and 238 are U.S. GOVERNMENT drawn or commissioned from http://search.usa.gov. See each caption for exact url.

Maps on pages 129 and 130 are from *CLOSING IN: MARINES IN THE SEIZURE OF IWO JIMA,* History & Museums Division, Headquarters Marine Corps, 1994. These illustrations are from pages 8 and 5 of that publication, respectively.

Map on page 131 is from *IWO JIMA: AMPHIBIOUS EPIC,* by Lt.Col. Whitman S. Bartley, based on sketches by 31st Naval Construction Battalion, Historical Branch, Headquarters Marine Corps, 1954. This illustration is from page 140 of that publication.

Map on page 133 is from "THE FINAL CAMPAIGN: MARINES IN THE VICTORY ON OKINAWA," by Colonel Joseph H. Alexander, *Marines in WWII Commemorative Series,* History & Museums Division, Headquarters Marine Corps, 1996. This illustration is from page 17 of that publication.

Pictures on pages 139, 140, and 142 are from "Army/Marine Clipart," U.S. AIR UNIV. (www.au.af.mil/au/awc/awcgate/cliparmy.htm). The second has image designator "1-07a.tif."

Pictures on pages 141, 160, 171, 228, and 229 reprinted with permission of Sorman Information and Media and the Swedish Armed Forces, from *SOLDF: SOLDATEN I FALT,* by Forsvarsmakten, with illustrations by Wolfgang Bartsch. These illustrations are from pages 225, 370, 220, 22/23, and 289 of the Swedish publication, respectively. Copyright © 2001 by Forsvarsmakten, Stockholm. All rights reserved.

Pictures on pages 158, 166, and 176 reprinted with permission of Dr. Anatol Taras, Minsk, Belarus, from *PODGOTOVKA RAZVEGCHIKA: SISTEMA SPETSNAZA GRU,* by A.E. Taras and F.D. Zaruz. These illustrations are from pages 173, 144/145/152, and 373 of the Belarus publication, respectively. Copyright © 1998 by A.E. Taras and F.D. Zaruz. All rights reserved.

Picture on page 158 reprinted with permission of Paladin Press, Boulder, CO, from *TACTICAL TRACKING OPERATIONS,* by David Scott-Donelan. This illustration ("Effects of Age and Weather on Spoor") is from page 42 of that publication. Copyright © 1998 by David Scott-Donelan. All rights reserved.

Picture on page 200 reprinted courtesy of EDWARD MOLINA. Copyright © 2012 by Edward Molina. All rights reserved.

Picture on page 212 is from *A CONCISE HISTORY OF THE UNITED STATES MARINE CORPS 1775-1969,* by Capt. William D. Parker, with sketch by Capt. Donald L. Dickson, History Division, Headquarters Marine Corps, 1970. This illustration is from page 62 of that publication.

Text:

ENDNOTES

Preface

1. Gen. Charles C. Krulak, "The Strategic Corporal: Leadership in the Three Block War," *Marines Magazine,* January 1999.
2. Memorandum for the record by H.J. Poole.

Chapter 1: *Evil Has Long Influenced War*

1. *The Jerusalem Bible,* Matthew 5:44 and 22:21, and Luke 6:26; "Sergeant Alvin York," by Dr. Michael Birdwell, Great War Society, as retrieved on 15 October 2009 from its website, www.worldwar1.com; "Sergeant York," DVD, 134 minutes, Warner Brothers Pictures, isbn #1-4198-3829-6.
2. President Ronald Reagan, "Message to the Senate Transmitting a Protocol to the 1949 Geneva Conventions," as retrieved from this url: http://www.reagan.utexas.edu/archives/speeches/1987/012987b.htm; Legal Information Inst., "Geneva Conventions," as retrieved from the following url: http://www.law.cornell.edu/wex/geneva_conventions; *Wikipedia Encyclopedia,* s.v. "Protocol I" and "Protocol II."
3. James M. McPherson, *Battle Cry of Freedom* (New York: Oxford Univ. Press, 1988), p. 580; *Wikipedia Encyclopedia,* s.v. "Battle of Stone's River."
4. Pope John Paul II, *Crossing the Threshold of Hope* (New York: Alfred A. Knopf, 1995), pp. 205, 206.
5. "Stonewall Jackson" (n.p, n.d.), p. 326, as retrieved from this url: www.sonofthesouth.net/leefoundation/jackson/battle-fredericksburg.html
6. "Conventional Weapon: 30/11/2011 Overview," Internat. Committee for the Red Cross, from the following url on 28 March 2014: http://www.icrc.org/eng/war-and-law/weapons/conventional-weapons/ overview-conventional-weapons.htm.
7. Thomas H. Green, *Weeds among the Wheat* (Notre Dame, IN: Ave Maria Press, 1984), pp. 28, 29 (for Christianity); "Satan and Humanity," as retrieved from the following url in December 2012, http://www.islamawareness.net/Jinn/satan.html (for Islam); *Torah (Old Testament),* 1 Chronicles 21:1 and Job 1:11 (for Judaism); *Wikipedia Encyclopedia,* s.v. "Devil" (for Hinduism and others).
8. "Netherworld," as retrieved from the following url: http://chinesemythical.wordpress.com.
9. *The Jerusalem Bible,* Ephesians, 6:10-12.
10. Ibid., Hebrews 12:22 with Revelations 9:1 and 12:3-9; *Gotquestions.org.* s.v. "Did one third of the angels fall with Lucifer?"
11. *The Jerusalem Bible,* 2 Corinthians, 12:8.

12. Ibid., 2 Kings, chapt. 6.
13. *About.com,* s.v. "Angels & Miracles."
14. "The Angels of Mons," *Great Mysteries of the 20th Century* (Pleasantville, NY: The Reader's Digest Assn., 1999), p. 114.
15. *Www.japanesebushido.org,* s.v. "Dictionary entry."
16. Memorandum for the record by H.J. Poole.
17. "From Makin to Bougainville: Marine Raiders in the Pacific War," by Maj. Jon T. Hoffman, *Marines in WWII Commemorative Series* (Washington, D.C.: Marine Corps Hist. Ctr., 1995), pp. 1-5.
18. "The Pacific," DVD, 10-part miniseries, 530 minutes, from Tom Hanks, Steven Spielberg, and Gary Goetzman, HBO and DreamWorks, n.d.
19. *The Quotations Page,* s.v. "Edmund Burke," from the following url: www.quotationspage.com/quotes/Edmund_Burke.

Chapter 2: *Mystical Tricks of a More Worldly Origin*

1. *The Travels of Marco Polo,* revised from Marsden's trans., ed. Manuel Komroff (New York: Modern Library, 1953), pp. 44, 45.
2. Editor's footnote, *The Travels of Marco Polo,* revised from Marsden's trans., ed. and intro. by Manuel Komroff (New York: Liveright Pub. Corp., 1926), p. 45.
3. Documentary of house examination by supernatural investigators, videotape, as viewed by author on educational TV, n.p., n.d.; *Wikipedia Encyclopedia,* s.v. "Ectoplasm."
4. Paul Begg, "Into Thin Air," in no. 31, vol. 3, *The Unexplained: Mysteries of Mind Space & Time,* from Mysteries of the Unexplained Series (Pleasantville, NY: Readers Digest Assn., 1992), p. 125.
5. *The Shadow,* audio CD, "The Making of a Legend," GAA Corp., 1996. stock#49030.
6. Dr. Masaaki Hatsumi, *The Essence of Ninjutsu* (Chicago: Contemporary Books, 1988), pp. 12, 55; Stephen K. Hayes, *Ninjutsu: The Art of the Invisible Warrior* (Chicago: Contemporary Books, 1984), p. 151.
7. Stephen K. Hayes, *Ninjutsu: The Art of the Invisible Warrior* (Chicago: Contemporary Books, 1984), p. 151.
8. Hatsumi, *The Essence of Ninjutsu,* pp. 33, 39.
9. Dr. Masaaki Hatsumi, *Ninjutsu: History and Tradition* (Burbank, CA: Unique Pubs., 1981), p. 7; Hatsumi, *The Essence of Ninjutsu,* pp. 101, 102; Ashida Kim, *The Invisible Ninja: Ancient Secrets of Surprise* (New York: Citadel Press, 1983), p. 13.
10. Stephen K. Hayes, *Legacy of the Night Warrior* (Santa Clarita, CA: Ohara Pubs., 1985), p. 26.
11. Hayes, *Ninjutsu: The Art of the Invisible Warrior,* p. 153.

12. Dr. David H. Reinke (expert on parapsychology and Eastern religions), in telephone conversation with author in June 2001.

13. Stephen K. Hayes, *The Mystic Arts of the Ninja: Hypnotism, Invisibility, and Weaponry* (Chicago: Contemporary Books, 1985), p. 102.

14. Ashida Kim, *The Invisible Ninja: Ancient Secrets of Surprise* (New York: Citadel Press, 1983), pp. 1, 56.

15. Hayes, *The Mystic Arts of the Ninja*, p. 1.

16. Hatsumi, *Ninjutsu: History and Tradition*, p. 3.

17. Ibid., author's preface page.

18. Hayes, *Legacy of the Night Warrior*, p. 20; Hatsumi, *Ninjutsu: History and Tradition*, p. 181.

19. Hayes, *Ninjutsu: The Art of the Invisible Warrior*, p. 153.

20. Ibid.

21. Hayes, *The Mystic Arts of the Ninja*, p. 1.

22. Hatsumi, *Ninjutsu: History and Tradition*, p. 5.

23. Ibid., p. 14.

24. Hatsumi, *The Essence of Ninjutsu*, p. 39.

25. Hatsumi, *Ninjutsu: History and Tradition*, p. 13; Hayes, *Ninjutsu: The Art of the Invisible Warrior*, p. 154.

26. Hayes, *The Mystic Arts of the Ninja*, p. 136.

27. Hatsumi, *Ninjutsu: History and Tradition*, p. 12.

28. Hayes, *The Mystic Arts of the Ninja*, pp. 132-138.

29. Ibid., p. 134.

30. Kim, *Secrets of the Ninja*, pp. 5-31.

31. Hatsumi, *Ninjutsu: History and Tradition*, author's preface.

32. Hayes, *Ninjutsu: The Art of the Invisible Warrior*, p. 154.

33. Hayes, *The Mystic Arts of the Ninja*, p. 137.

34. Hayes, *Ninjutsu: The Art of the Invisible Warrior*, p. 157.

35. Ibid.

36. Ibid.

37. Ibid.

38. Hayes, *The Mystic Arts of the Ninja*, p. 139.

39. Kim, *The Invisible Ninja*, pp. 1, 56; Kim, *Secrets of the Ninja*, pp. 10-29; Hayes, *The Mystic Arts of the Ninja*, pp. 133-135.

40. Kim, *The Invisible Ninja*, p. 56.

41. Hayes, *The Mystic Arts of the Ninja*, p. 135; Kim, *Secrets of the Ninja*, pp. 10-30.

42. Hatsumi, *The Essence of Ninjutsu*, p. 39.

43. Hayes, *The Mystic Arts of the Ninja*, p. 133; Hatsumi, *Ninjutsu: History and Tradition*, pp. 231, 232.

44. Hatsumi, *The Essence of Ninjutsu*, p. 40.

45. Kim, *Secrets of the Ninja*, p. 83.

46. Kim, *The Invisible Ninja*, p. 13.

47. Hayes, *The Mystic Arts of the Ninja*, p. 102.

48. Kim, *Secrets of the Ninja*, p. 95.

49. Ibid., p. 149.
50. Ibid.
51. Ibid.
52. Ibid., p. 150.
53. Hayes, *Legacy of the Night Warrior,* p. 154.
54. "Hypnosis," in *The Complete Manual of Fitness and Well-Being* (Pleasantville, NY: The Reader's Digest Assn., 1984), p. 330.
55. Ibid.
56. *Night Movements*, trans. and preface by C. Burnett (Tokyo: Imperial Japanese Army, 1913; reprint (Port Townsend, WA: Loompanics Unlimited, n.d.), p. 69. (This work will be henceforth cited as *Night Movements.)*
57. Reinke, Dr. David H. (expert on parapsychology and Eastern religions), in series of e-mails after June 2011.
58. Hatsumi, *The Essence of Ninjutsu,* p. 39.
59. Hayes, *The Mystic Arts of the Ninja,* p. 135.
60. Ashida Kim, *Secrets of the Ninja* (New York: Citadel Press, 1981), p. 10.

Chapter 3: *Overwhelming Force No Longer the Answer*

1. *Okinawa: The Last Battle,* by Roy E. Appleman, James M. Burns, Russell A. Gugeler, and John Stevens, *United States Army in World War II Series* (Washington, D.C.: U.S. Army's Ctr. of Mil. Hist., 2000), pp. 322, 323.
2. H. John Poole, *Gung Ho: The Corps' Most Progressive Tradition* (Emerald Isle, NC: Posterity Press, 2012), pp. 140, 141.
3. "The Final Campaign: Marines in the Victory on Okinawa," by Joseph H. Alexander, *Marines in WWII Commemorative Series* (Washington, D.C.: Hist. and Museums Div., HQMC, 1996), pp. 41-52.
4. Poole, *Gung Ho,* pp. 141-145.
5. *Tennozan: The Battle of Okinawa and the Atomic Bomb,* by George Feifer (Boston: Houghton Miflin, 1992), p. 578, and *Okinawa: The Last Battle,* by Appleman et al, p. 468, in *Military History Online,* s.v. "Battle of Okinawa."
6. Thomas M. Huber, "Japan's Battle for Okinawa, April - June 1945," *Leavenworth Papers No. 18* (Ft. Leavenworth, KS: Combat Studies Inst., U.S. Army's Cmd. & Gen. Staff College, 1990).
7. Col. David H. Hackworth U.S. Army (Ret.) and Julie Sherman, *About Face* (New York: Simon & Schuster, 1989), p. 594.
8. Shaun Waterman, "North Korean Jamming of GPS Shows System's Weakness," *Washington Times,* 23 August 2012; "Chinese GPS Killers," *Strategy Page,* 23 November 2007.

9. Lt.Col. R.F. Reid-Daly, *Pamwe Chete: The Legend of the Selous Scouts* (Weltevreden Park, South Africa: Covos-Day Books, 1999), p. 182.
10. H. John Poole, *Terrorist Trail: Backtracking the Foreign Fighter* (Emerald Isle, NC: Posterity Press, 2006), p. 152.
11. Joseph S. Bermudez, Jr., *North Korean Special Forces* (Annapolis, MD: Naval Inst. Press, 1998).
12. H. John Poole, *Tequila Junction: 4th-Generation Counterinsurgency* (Emerald Isle, NC: Posterity Press, 2008), pp. 216-219.

Chapter 4: *Enhancement of Local Security More Vital*

1. Peter Stiff, *The Silent War: South African Recce Operations, 1969-1994* (Alberton, South Africa: Galago Pub., 1999), pp. 289, 290.
2. Joseph J. Collins, "Counterinsurgency & Common Sense," *Armed Forces Journal,* January/February 2013, p. 17.
3. Don Moser, "Their Mission Defend, Befriend," *Life Magazine,* 25 August 1967.
4. Poole, *Gung Ho,* chapt. 14.
5. *Afghanistan Online* (www.afghan-web.com), s.v. "History Chronology"; *Wikipedia Encyclopedia,* s.v. "Afghanistan," "Ptolemaic Dynasty," "Kingdom of Seleucus," "Persian Empire," "Alexander the Great," "Mongol Empire," "Timur," and "Mughal Empire."
6. Pauline Jelinek and Anne Gearan (AP), "General Seeks New Afghan Approach," *Philadelphia Inquirer,* 1 August 2009.
7. Ibid.
8. Collins, "Counterinsurgency & Common Sense," p. 19.

Chapter 5: *The Pentagon's New Worldwide Strategy*

1. President Obama's strategy announcement, ABC's Nightly News, 12 January 2010.
2. Carlos Munoz, "Pentagon Taking New Tact in Terrorism Fight in Africa," *The Hill,* 14 July 2012.
3. "Joseph Kony Hunt is Proving Difficult for U.S. Troops," *Washington Post,* 29 April 2012.
4. Poole, *Terrorist Trail,* chapt. 10.
5. "Joseph Kony Hunt is Proving Difficult for U.S. Troops."
6. Ibid.
7. Ibid.

8. NPR's "Morning Edition" News, 23 February 2009; "Wushu & Sanda," *Fight Quest,* Discovery Channel, 28 December 2008.
9. "China-U.S. Relations: Current Issues and Implications for U.S. Policy," by Kerry Dumbaugh, Congressional Research Service (CRS) Report for Congress, Order Code RL33877 (Washington, D.C.: Library of Congress, 1 October 2007); *Wikipedia Encyclopedia,* s.v. "Haiti."
10. Loveday Morris, "In Syria, Hezbollah Forces Appear Ready to Attack Rebels in City of Aleppo," *Washington Post,* 2 June 2013; CBS's "Nightly News" (one *Sepah* squad per neighborhood in Allepo segment), 30 October 2013.
11. Robin Wright, *Sacred Rage: The Wrath of Militant Islam* (New York: Simon & Schuster, 1985), pp. 33-35, as quoted in *Warriors of Islam: Iran's Revolutionary Guard,* by Kenneth Katzman (Boulder, CO: Westview Press, 1993), p. 71; Sepehr Zabih, *The Iranian Military in Revolution and War* (London: Routledge, 1988), pp. 210-212.

Chapter 6: *How Best to Train Local Security Forces*

1. Multi-tour U.S. Army Special Forces veteran of wars in Iraq and Afghanistan and expert mantracker, in e-mail conversation with author on 5 June 2013.
2. Dr. Haha Lung, *Knights of Darkness: Secrets of the World's Deadliest Night Fighters* (Boulder, CO: Paladin Press, 1998), pp. 8, 9.
3. *Night Movements*; H. John Poole, *Dragon Days: Time for "Unconventional" Tactics* (Emerald Isle, NC: Posterity Press, 2007), chapts. 16 and 17.
4. "Handbook on the Chinese Communist Army," *DA Pamphlet 30-51* (Washington, D.C.: Hdqts. Dept. of the Army, 1960), p. 29.
5. Bermudez, *North Korean Special Forces,* pp. 233-248.
6. Memorandum for the record by H.J. Poole.
7. H. John Poole, *The Tiger's Way: A U.S. Private's Best Chance for Survival* (Emerald Isle, NC: Posterity Press, 2003), pp. 255-266.
8. Curator of Iranian War Museum at Tehran, in conversation with author around 2002.
9. Dan Murphy, "Sadr the Agitator: Like Father, Like Son," *Christian Science Monitor,* 27 April 2004, p. 6.
10. H. John Poole, *Tactics of the Crescent Moon* (Emerald Isle, NC: Posterity Press, 2004), p. 208.
11. Brigadier Mohammad Yousaf and Maj. Mark Adkin, *Bear Trap: Afghanistan's Untold Story* (South Yorkshire, UK: Leo Cooper, n.d.).
12. H. John Poole, *Phantom Soldier: The Enemy's Answer to U.S. Firepower* (Emerald Isle, NC: Posterity Press, 2001), p. 135.
13. Yousaf and Adkin, *Bear Trap.*
14. Ibid.

15. Ibid.

16. Former Marine twin brother of CWO-4 in charge of indigenous Afghan military training, in conversation with author around 2011.

17. Mao Tse-tung, "Mao's Primer on Guerrilla War," trans. B.Gen. Samuel B. Griffith, in FMFRP 19-9, *The Guerrilla and How to Fight Him* (Quantico, VA: Marine Corps Combat Develop. Cmd., 1990), p. 7, and in *Marine Corps Gazette*, January 1962 (plus a 1941 issue).

18. "Handbook On U.S.S.R. Military Forces," *TM 30-340,* p. V-104.

19. Lt.Col. David M. Glantz, Curriculum Supervisor, in foreword to "Soviet Night Operations in World War II," by Maj. Claude R. Sasso, *Leavenworth Papers No. 6* (Ft. Leavenworth, KS: Combat Studies Inst., U.S. Army's Cmd. & Gen. Staff College, 1982), p. viii.

20. *Soviet Combat Regulations of November 1942* (Moscow: [Stalin], 1942), republished as *Soviet Infantry Tactics in World War II: Red Army Infantry Tactics from Squad to Rifle Company from the Combat Regulations,* with trans., intro., and notes by Charles C. Sharp (West Chester, OH: George Nafziger, 1998), republisher's intro., p. 3.

21. Maj. Claude R. Sasso, "Soviet Night Operations in World War II," *Leavenworth Papers No. 6* (Ft. Leavenworth, KS: Combat Studies Inst., U.S. Army's Cmd. & Gen. Staff College, 1982), p. 35.

22. *The Bear Went over the Mountain: Soviet Combat Tactics in Afghanistan,* trans. and ed. Lester W. Grau, Foreign Mil. Studies Office, DoD (Soviet Union: Frunze Mil. Academy, n.d.; reprint Washington, D.C.: Nat. Defense Univ. Press, 1996), p. 38. (This work will henceforth be cited as *The Bear Went over the Mountain.)*

23. Editor's commentary, *The Bear Went over the Mountain,* p. 174.

24. Poole, *The Tiger's Way,* fig. 7.2.

25. *Soviet Combat Regulations of November 1942*, p. 22.

26. Ibid., p. 51.

27. Jim Simpson, "Scouts to the Rescue," *Defense Watch,* 17 September 2003.

Chapter 7: *Light Infantrymen Need No Technology*

1. Memorandum for the record by H.J. Poole; *Wikipedia Encyclopedia,* s.v. "Protocol I" and "Protocol II."

2. Poole, *The Tiger's Way,* pp. 79-85.

3. Poole. *Phantom Soldier,* pp. 33, 34.

4. Memorandum for the record by H.J. Poole.

5. Ibid.

6. Ibid.

7. Ibid.

8. Tag Guthrie (former squad leader in A/1/4), in e-mail to the author during 2012.

9. Memorandum for the record by H.J. Poole.

10. Bruce I. Gudmundsson, *Stormtroop Tactics—Innovation in the German Army 1914-1918* (New York: Praeger, 1989).

11. Memorandum for the record by H.J. Poole.

12. Maj.Gen. Robert H. Scales, U.S. Army (Ret.), "Infantry and National Priorities," *Armed Forces Journal,* December 2007, pp. 14-17.

13. Ibid.

14. "Why Did Armored Corps Fail in Lebanon," by Hanan Greenberg, Israeli News, 30 August 2006.

Chapter 8: *Taking Strongpoints without Bombardment*

1. H. John Poole, *One More Bridge to Cross: Lowering the Cost of War* (Emerald Isle, NC): Posterity Press, 1999), pp. 73-75.

2. Chesty Puller, in *Marine,* by Burke Davis (New York: Bantam, 1964), p. 390.

3. Gen. Vo Nguyen Giap, "Once Again We Will Win," as quoted in *The Military Art of People's War,* ed. Russel Stetler (New York: Monthly Review Press, 1970), pp. 264, 265.

4. Memorandum for the record by H.J. Poole.

5. Ibid.

6. Origins at url: http://deadliestfiction.wikia.com/wiki/Dac_Cong.

7. Hackworth and Sherman, *About Face,* p. 594.

8. *Isolation of Rabaul,* by Henry I. Shaw and Maj. Douglas T. Kane USMC, part III, chapt. 3, "Assault on Cape Torokina," vol. II, *History of U.S. Marine Corps Operations in World War II Series* (Washington, D.C.: Hist. Branch, HQMC, 1963), pp. 211-213.

9. Kelley Navy Cross citation, at Assn. of 3rd Battalion, 4th Marines website.

Chapter 9: *Monitoring a Large Area with Few U.S. Forces*

1. Kirk Hauser (one of Sgt. Price's men), in e-mails to author in December 2013; Sgt. Toney Price, sketch of trail junction ambush, n.d.

2. Sgt. Toney Price, two-hour interview by Kirk Hauser, CD, 2013.

3. Ibid.

4. *Wikipedia Encyclopedia,* s.v. "Tony Stein."

5. "We Were Soldiers," DVD, 84 Minutes, Warner Brothers, 2002, based on the book, *We Were Soldiers Once and Young: Ia Drang—the Battle That Changed the War in Vietnam,* by Harold G. Moore and Joseph Galloway (New York: Presidio Press, 2004).

6. Sgt. Toney Price, two-hour interview by Kirk Hauser, CD, 2013.

7. Bill Gertz, "Notes from the Pentagon," *Washington Times,* 5 March 2004; NPR's Morning News, 1 May 2006; Poole, *Terrorist Trail,* pp. 25, 26.
8. Terrence Maitland and Peter McInerney, *Vietnam Experience: A Contagion of War* (Newton, MA: Boston Pub., 1968), p. 94.
9. Poole, *Dragon Days,* pp. 120, 121.

Chapter 10: *New Doctrine Calls for More Squad Autonomy*

1. "Windtalkers," TV movie, 134 minutes, MGM, 2002.
2. Brad Bennett (81mm operator at CAC 10), in e-mail to author on 18 February 2013.
3. Ron Baldwin (member of CAC 10), in Amazon review of *Gung Ho,* 8 January 2013.
4. *Tactics,* FMFM 1-3 (Washington, D.C.: HQMC, 1991), foreword; "Tactical Fundamentals," MCI 7401 (Washington, D.C.: Marine Corps Inst., n.d.), p. 43.
5. William S. Lind, *Maneuver Warfare Handbook* (Boulder, CO: Westview Press, 1985), p. 2.
6. Ibid., p. 12.
7. Poole, *Phantom Soldier,* chapt. 7.
8. Lind, *Maneuver Warfare Handbook,* p. 25.
9. William S. Lind, "The Theory and Practice of Maneuver Warfare," *Maneuver Warfare: An Anthology,* ed. Richard D. Hooker, Jr. (Novato, CA: Presidio Press, 1993), p. 10.
10. Ibid., p. 7.
11. Gudmundsson, *Stormtroop Tactics,* pp. 146, 147.
12. Timothy T. Lupfer, "The Dynamics of Doctrine: The Changes in German Tactical Doctrine during the First World War," *Leavenworth Papers No. 4* (Fort Leavenworth, KS: Combat Studies Inst., U.S. Army's Cmd. & Gen. Staff College, 1981), in MCI 7401, *Tactical Fundamentals,* 1st course of Warfighting Skills Program (Washington, D.C.: Marine Corps Inst., 1989), p. 43.
13. Poole, *Gung Ho,* appendix B.
14. *Tactics,* FMFM 1-3, foreword.
15. Poole, *Gung Ho,* chapts. 5 and 6.

Chapter 11: *Riflemen Need More Than Rules of Engagement*

1. "From Makin to Bougainville," by Hoffman, pp. 1-5.
2. Ibid., p. 5.

3. *Gunny G.'s Globe and Anchor,* s.v. "gung ho."

4. Bruce Knipp, "Letter to the Editor," *Jacksonville Daily News* (NC), 3 December 2013.

5. Ibid.

6. Ibid.

7. Ibid.

8. U.S. Navy Chaplains' Creed as remembered by author.

9. James Corbett, *Jim Corbett's India,* ed. R.E. Hawkins (London: Oxford Univ. Press, 1978), p. 25.

10. James Corbett, *Jungle Lore* (London: Oxford Univ. Press, 1953), p. 172.

11. About.com, s.v. "Angels and Miracles," from the following urls: http://angels.about.com/od/AngelsReligiousTexts/f/What-Are-Some-Muslim-Guardian-Angel-Prayers.htm, and http://angels.about.com/od/AngelBasics/f/How-Do-Guardian-Angels-Guide-Me.htm

12. Hatsumi, *Ninjutsu: History and Tradition,* p. 12

13. Ibid., p. 3.

14. Ibid., author's preface page.

15. Memorandum for the record by H.J. Poole.

16. Poole, *Gung Ho,* chaps. 2 and 3.

17. Sun Tzu, *The Art of War,* trans. and with intro. by Samuel B. Griffith, foreword by B.H. Liddell Hart (New York: Oxford Univ. Press, 1963), p. 82.

18. Ibid., p. 84.

Chapter 12: *Creating a More Proactive Fighter*

1. *Wikipedia Encyclopedia,* s.v. "Clyde A. Thomason."

2. Ibid.

3. Mao Tse-tung, "Mao's Primer on Guerrilla War," trans. B.Gen. Samuel B. Griffith.

4. *Wikipedia Encyclopedia,* s.v. "Clyde A. Thomason."

5. *Dan Marsh's Marine Raider Page* (USMCRaiders.com), created by a former member of 4th Raider Battalion, home page entry.

6. *Wikipedia Encyclopedia,* s.v. "PFC Henry Gurke."

7. Kim, *Secrets of the Ninja,* pp. 5-31.

8. Hatsumi, *Ninjutsu: History and Tradition,* author's preface.

9. Hayes, *Ninjutsu: The Art of the Invisible Warrior,* p. 154.

10. Hayes, *The Mystic Arts of the Ninja,* p. 137.

11. Hayes, *Ninjutsu: The Art of the Invisible Warrior,* p. 157.

12. Ibid.

13. Ibid.

14. Ibid.

15. Hayes, *The Mystic Arts of the Ninja,* p. 139.

16. Memorandum for the record by H.J. Poole.
17. "Brave Heart," VHS, 177 minutes, Paramount, 1995.
18. Poole, *The Tiger's Way,* part two.
19. "Gung Ho: The Story of Carlson's Makin Island Raiders," movie, 87 minutes, Universal Studios, 1943.
20. Poole, *Gung Ho,* chapt. 7.
21. Giap, "Once Again We Will Win," pp. 264, 265.
22. Poole, *The Tiger's Way,* table 2.2.

Chapter 13: *What Such Riflemen Add to Unit Power*

1. *Wikipedia Encyclopedia,* s.v. "William Gary Walsh."
2. Lt.Col. Whitman S. Bartley, *Iwo Jima: Amphibious Epic* (Washington, D.C.: Hist. Branch, HQMC, 1954), p. 164.
3. Ibid., p. 149.
4. Bill D. Ross, *Iwo Jima: Legacy of Valor* (New York: Vintage, 1986), p. 249.
5. Bartley, *Iwo Jima,* pp. 140, 141.
6. *Wikipedia Encyclopedia,* s.v. "William Gary Walsh."
7. Ibid., s.v. "Richard Earl Bush"; Louie (Dan Marsh's son) at USMCRaiders.com in e-mail to author of 9 December 2013.
8. John Wukovits, *American Commando* (New York: New American Library, 2009), p. 61.
9. *Dan Marsh's Marine Raider Page* (USMCRaiders.com), s.v. "Okinawa: Mt. Yae Take."
10. *Wikipedia Encyclopedia,* s.v. "Richard Earl Bush."
11. "Finland in the Winter War: An Examination of Finnish Infantry Tactics," by Ville Savin (n.p., n.d.).
12. Michael Lee Lanning and Dan Cragg, *Inside the VC and the NVA: The Real Story of North Vietnam's Armed Forces* (New York: Ivy Books, 1992), pp. 211, 212.

Chapter 14: *All-Hands Accountability for More Moral Units*

1. *Wikipedia Encyclopedia,* s.v. "List of MoH Recipients for WWII," "List of Korean War MoH Recipients," and "List of MoH Recipients for the Vietnam War."
2. Memorandum for the record by H.J. Poole.
3. *Handbook on the Chinese Armed Forces,* DDI-2680-32-76 (Washington, D.C.: DIA, July 1976), p. ii.
4. Ibid., pp. 5-27, 5-28.
5. Truong Chinh, *Primer for Revolt,* intro. Bernard B. Fall (New York: Praeger, 1963), pp. 114-117.

6. *Handbook on the Chinese Armed Forces,* DDI-2680-32-76, pp. 5-19.

7. Ibid., p. 5-28.

8. Poole, *Gung Ho,* chapt. 3.

9. Maitland and McInerney, *Vietnam Experience: A Contagion of War,* p. 97.

10. "Gung Ho," Universal Studios.

11. Maj.Gen. Edwin Simmons USMC (Ret.), in "An Entirely New War" segment of *Korea — the Unknown War Series* (London: Thames TV in association with WGBH Boston, 1990), NC Public TV; Maj. Scott R. McMichael, "The Chinese Communist Forces in Korea," chapt. 2 of "A Historical Perspective on Light Infantry," *Leavenworth Research Survey No. 6* (Ft. Leavenworth, KS: Combat Studies Inst., U.S. Army's Cmd. & Gen. Staff College, 1987), p. 56.

12. Poole, *Gung Ho,* chapt. 5.

Chapter 15: *Reestablishing Individual Initiative*

1. Memorandum for the record by H.J. Poole.

2. *Wikipedia Encyclopedia,* s.v. "Private Carlton W. Barrett."

3. Ibid., s.v. "John J. Pinder, Jr."

4. Ibid., s.v. "Private Carlton W. Barrett."

5. Jennifer Griffin, "U.S. Intelligence Officer Reveals Secret Story of Saddam Hussein's Capture," FoxNews.com, 12 December 2008.

6. Col. Merritt A. Edson, in *Fighting on Guadalcanal* (Washington, D.C.: U.S.A. War Office, 1942), pp. 14-19.

7. Dutch Marine exchange officer in conversation with author on 16 January 2002.

8. "Gung Ho," Universal Studios.

9. Wukovits, *American Commando,* p. 53.

10. *China Page,* s.v. "Gung ho."

11. *The Strategic Advantage: Sun Zi & Western Approaches to War,* ed. Cao Shan (Beijing: New World Press, 1997), pp. 21, 22.

12. Poole, *Gung Ho,* chapt. 4.

13. Memorandum for the record by H.J. Poole.

14. "Images for *Dac Cong,*" as retrieved from this url in March 2014: http://www.google.com/images?oe=UTF-8&gfns=1&q=dac+cong&hl=en&sa=X&oi=image_result_group&ei=r281U57HKbPQsQTqxIGgCg&ved=0CBsQsAQ.

15. Kirkland C. Peterson, *Mind of the Ninja: Exploring the Inner Power* (Chicago: Contemporary Books, 1986), p. 105.

16. Hayes, *Legacy of the Night Warrior,* p. 152.

17. *U.S. Special Forces Reconnaissance Manual* (n.p., n.d.); reprint, Sims, AR: Lancer Militaria, 1982), p. 51.

18. "From Makin to Bougainville," by Hoffman, p. 23; Poole, *Gung Ho,* p. 83.

Chapter 16: *Personal-Decision-Making Practice*

1. Poole, *Gung Ho,* pp. 138-145.
2. "Finland in the Winter War: An Examination of Finnish Infantry Tactics," by Ville Savin (n.p., n.d.).
3. Memorandum for the record by H.J. Poole.

Chapter 17: *Troops Must Help to Design Own Moves*

1. "Sands of Iwo Jima," videocassette, 109 min., Republic Pictures, 1988, Tarawa portion.
2. Ibid.
3. *Wikipedia Encyclopedia,* s.v. "William D. Hawkins."
4. *Wikipedia Encyclopedia,* s.v. "Tony Stein."
5. Poole, *Gung Ho,* chaps. 9 and 10.
6. Ibid., chaps. 11 and 13.
7. Memorandum for the record by H.J. Poole.
8. Poole, *Gung Ho,* appendix B.
9. Wukovits, *American Commando,* p. 54.
10. Ibid., p. 53.
11. Ibid., p. 56.
12. Poole, *Gung Ho,* p. 33.
13. *Phrases.org,* s.v. "Gung ho," quote from *New York Times,* 1942.
14. Lt.Col. Gregory H. Kitchens USMCR, "Building the Team for Unit Excellence," *Marine Corps Gazette,* December 2003.
15. Jim Schmorde (Vietnam veteran of L/3/4), in e-mail to author of December 2012.
16. Memorandum for the record by H.J. Poole.
17. Wukovits, *American Commando,* p. 12.
18. Poole, *The Tiger's Way*, part three.
19. "Criminal Minds," *J.J. Episode,* ION TV, 9 December 2013.
20. Jesus, as quoted in *The Jerusalem Bible,* John, 14:6.

Afterword: *No Minor Oversight*

1. Seal Team Six member, in interview on CBS's "60 Minutes," n.d.

2. Highly experienced reconnaissance Marine who after retirement became training contractor for special-operations community, in an e-mail to author during January 2014.
3. Memorandum for the record by H.J. Poole.
4. Gudmundsson, *Stormtroop Tactics—Innovation in the German Army 1914-1918,* p. 94.
5. Nicholas Warr, *Phase Line Green: The Battle for Hue, 1968* (Annapolis, MD: Naval Inst. Press, 1997), pp. 155-160.

Appendix: *Korean War Sighting*

1. *TFP Student Action* website (www.tfpstudentaction.org), s.v. "Michael, Michael of the Morning Prayer."
2. "Michael, Michael of the Morning: A Young Marine's Letter to His Mom," *The Remnant Newspaper* (Forest Lake, MN), 25 December 2008.
3. Ibid.

Glossary

AK-47	Small-arms designator	Communist Bloc assault rifle
ARVN	Army of the Republic of Vietnam	South Vietnamese army
BAR	Browning Automatic Rifle	Automatic weapon for WWII U.S. infantry squad
CAC	Combined Action Company	Initial term in Vietnam for U.S. Marine squads serving with local militia platoons
CAP	Combined Action Platoon	Eventual term when one fire team of Marines assigned to each of three militia squads
CCW	Certain Conventional Weapons	Geneva Convention term
CO	Commanding Officer	Individual with command authority over a military unit
COIN	Counterinsurgency	How to combat guerrillas
COPD	Chronic Obstructive Pulmonary Disease	Progressively more serious breathing problems
CP	Command Post	Location of unit commander
DIA	Defense Intelligence Agency	American military espionage organization
DMZ	Demilitarized Zone	Unfortified frontier buffer between nations
DoD	Department of Defense	Civilian headquarters of all U.S. military services

E-1	Pay grade	Lowest in the U.S. military
E-2	Pay grade	Second lowest in U.S. military
E-3	Pay grade	Third lowest in U.S. military
E-4	Pay grade	Fourth lowest in U.S. military
E-5	Pay grade	Fifth lowest in U.S. Military
E&E	Escape and Evasion	Art of eluding a pursuer
EEG	Electroencephalogram	Electronic mind study
FMCR	Fleet Marine Corps Reserve	Status of active-duty retirees until reaching 30-year mark
FMFM	Fleet Marine Field Manual	Military "how-to" book
4GW	Fourth-Generation Warfare	War waged in four arenas at once—religious/psychological, economic, political, martial
GI	Government Issue	Colloquial term for U.S. military service member
GPS	Global Positioning System	Location finder using satellite signals
GWOT	Global War on Terrorism	U.S. effort to defeat Al-Qaeda
HE	High Explosive	Type of munition
HQMC	Headquarters Marine Corps	Various staff sections through which U.S. Commandant manages the organization
INTER-POL	International Criminal Police Organization	Intergovernmental law enforcement agency shared by 190 countries
LRA	Lord's Resistance Army	Outlawed African militia that uses child soldiers
LZ	Landing Zone	Helicopter destination

M-16	Small-arms designator	U.S. assault rifle
M-60	Small-arms designator	U.S. machinegun
MoH	Medal of Honor	Highest award for bravery
MP	Military police	Military law enforcer
MW	Maneuver Warfare	Way of fighting where tactical surprise replaces firepower
NCO	Noncommissioned Officer	Military pay grades E-4 & E-5
NVA	North Vietnamese Army	U.S. foe during Vietnam War
NVG	Night Vision Goggles	Ambient-light device
OK	Okay	Word of approval
PF	Popular Forces	South Vietnamese militia
PFC	Private First Class	E-2 in Marines & E-3 in Army
PLA	People's Liberation Army	PRC's military establishment
POW	Prisoner of War	Captured enemy combatant
PRC	People's Republic of China	Mainland China
PT-76	Armor designator	Communist Bloc light amphibious tank
PTSD	Post Traumatic Stress Disorder	Mental problems after very troubling event
RL	Rocket launcher	Tube for firing tiny missiles
RPG	Rocket-Propelled Grenade	Personal anti-tank weapon
SEAL	Sea, Air, and Land	U.S. Navy commando entity specializing in overseas raids
2GW	Second-Generation Warfare	Focus on razing strongpoints
SIGINT	Signals Intelligence	Enemy message information

SOF	Special Operations Forces	Units capable of commando missions
SS	*Schutzstaffel*	WWII Nazi command accused of crimes against humanity
SWAT	Special-Weapons Assault Team	Paramilitary police squad
TAOR	Tactical Area of Responsibility	Assigned sector of operations
3GW	Third-Generation Warfare	Focus is on bypassing enemy's strongpoints to more easily destroy his strategic assets
U.N.	United Nations	Alliance of countries
U.S.	United States	America
USMC	United States Marine Corps	America's amphibious landing force in readiness
UW	Unconventional Warfare	Infantry aspects are E&E and how to fight like a guerrilla
VC	Viet Cong	NVA-trained/advised South Vietnamese guerrillas
WWI	World War I	First global conflict
WWII	World War II	Second global conflict
WWIII	World War III	Third global conflict
ZANLA	Zimbabwe African National Liberation Army	Communist rebel army in Rhodesia
ZANU P/F	Zimbabwe African National Union Patriotic Front	Political wing of Rhodesian Communist rebel army

Bibliography

U.S. Government Publications, Databases, and News Releases

Bartley, Lt.Col. Whitman S. *Iwo Jima: Amphibious Epic.* Washington, D.C.: History Branch, Headquarters Marine Corps, 1954.

Bougainville and the Northern Solomons. By Major John N. Rentz. "USMC Historical Monograph." Washington, D.C.: Historical Branch, Headquarters Marine Corps, 1946. As retrieved from url: ibiblio.org/hyperwar/USMC/USMC-M-NSols/USMC-M-NSol-2.html

Fighting on Guadalcanal. Washington, D.C.: U.S.A. War Office, 1942.

"The Final Campaign: Marines in the Victory on Okinawa." By Colonel Joseph H. Alexander. *Marines in WWII Commemorative Series.* Washington, D.C.: History and Museums Division, Headquarters Marine Corps, 1996.

"From Makin to Bougainville: Marine Raiders in the Pacific War." By Major Jon T. Hoffman. *Marines in WWII Commemorative Series.* Washington, D.C.: Marine Corps Historical Center, 1995.

Handbook on the Chinese Armed Forces. DDI-2680-32-76. Washington, D.C.: Defense Intelligence Agency, July 1976.

Huber, Thomas M. "Japan's Battle for Okinawa, April - June 1945." *Leavenworth Papers No. 18.* Ft. Leavenworth, KS: Combat Studies Institute, U.S. Army's Command and General Staff College, 1990. As retrieved on 19 November 2011 from the following url: www.cgsc.edu/carl/resources/csi/Huber/Huber.asp#contents.

Isolation of Rabaul. By Henry I. Shaw and Major Douglas T. Kane. *History of U.S. Marine Corps Operations in World War II Series.* Volume II. Washington, D.C.: Historical Branch, Headquarters Marine Corps, 1963. As retrieved from the following url: www.ibiblio.org/hyperwar/USMC/II/index.html#contents.

Lupfer, Timothy T. "The Dynamics of Doctrine: The Changes in German Tactical Doctrine during the First World War." *Leavenworth Papers No. 4.* Fort Leavenworth, KS: Combat Studies Institute, U.S. Army's Command and General Staff College, 1981. In MCI 7401, *Tactical Fundamentals,* 1st course of Warfighting Skills Program. Washington, D.C.: Marine Corps Institute, 1989.

McMichael, Major Scott R. "The Chinese Communist Forces
in Korea." *A Historical Perspective on Light Infantry*.
Chapter 2. *Leavenworth Research Survey No. 6*.
Fort Leavenworth, KS: Combat Studies Institute,
U.S. Army's Command and General Staff College, 1987.
Okinawa: The Last Battle. By Roy E. Appleman, James
M. Burns, Russell A. Gugeler, and John Stevens. *United
States Army in World War II Series*. Washington, D.C.:
U.S. Army's Center of Military History, 2000.
As retrieved on 30 November 2011 from the following url:
www.history.army.mil/books/wwii/okinawa/index.htm#contents.
"Tactical Fundamentals." MCI 7401. Washington, D.C.: Marine Corps
Institute, n.d.
Tactics. FMFM 1-3. Washington, D.C.: Headquarters Marine Corps,
1991.
U.S. Special Forces Reconnaissance Manual. N.p., n.d. Reprint.
Sims, AR: Lancer Militaria, 1982.

Civilian Publications

Analytical Studies, Databases, and Websites

About.com. As retrieved from its website.
Afghanistan Online. As retrieved from its website,
www.afghan-web.com.
"The Angels of Mons." *Great Mysteries of the 20th Century*.
Pleasantville, NY: The Reader's Digest Association, 1999.
Association of 3rd Battalion, 4th Marines' website. As retrieved from
this url: http://thundering-third.org.
Begg, Paul. "Into Thin Air." In issue 31, volume 3, *The Unexplained:
Mysteries of Mind Space & Time*, from *Mysteries of the
Unexplained*. Pleasantville, NY: Readers Digest Association, 1992.
Bermudez, Joseph S., Jr. *North Korean Special Forces*. Annapolis,
MD: Naval Institute Press, 1998.
Bhagavad Gita. By anonymous.
China Page. As retrieved from its website, www.chinapage.com.
Chinh, Truong. *Primer for Revolt*. Introduction by Bernard B. Fall.
New York: Praeger, 1963.
Corbett, James. *Jim Corbett's India*. Edited by R.E. Hawkins.
London: Oxford University Press, 1978.
Corbett, James. *Jungle Lore*. London: Oxford University Press, 1953.
Dan Marsh's Marine Raider Page (USMC.com). Created by a former
member of 4th Raider Battalion and now managed by his son,
Louie.

Davis, Burke. *Marine.* New York: Bantam, 1964.

Deadliest Fiction. As retrieved from its website,
http://deadliestfiction.wikia.com.

Got Questions. As retrieved from its website, www.gotquestions.org.

Green, Thomas H. *Weeds among the Wheat.* Notre Dame, IN:
Ave Maria Press, 1984.

Gudmundsson, Bruce I. *Stormtroop Tactics—Innovation
in the German Army 1914-1918.* New York: Praeger, 1989.

Gunny G.'s Globe and Anchor. From its website, www.angelfire.com.

Hackworth, David H. and Julie Sherman. *About Face.* New York:
Simon & Schuster, 1989.

Hatsumi, Dr. Masaaki. *The Essence of Ninjutsu.* Chicago:
Contemporary Books, 1988.

Hatsumi, Dr. Masaaki. *Ninjutsu: History and Tradition.* Burbank,
CA: Unique Publications, 1981.

Hayes, Stephen K. *Legacy of the Night Warrior.* Santa Clarita, CA:
Ohara Publications, 1985.

Hayes, Stephen K. *The Mystic Arts of the Ninja: Hypnotism,
Invisibility, and Weaponry.* Chicago: Contemporary Books, 1985.

Hayes, Stephen K. *Ninjutsu: The Art of the Invisible Warrior.*
Chicago: Contemporary Books, 1984.

"Hypnosis." In *The Complete Manual of Fitness and Well-Being.*
Pleasantville, NY: The Reader's Digest Association, 1984.

Www.japanesebushido.org. As retrieved from its website.

The Jerusalem Bible. Garden City, NY: Doubleday, 1966.

John Paul II, Pope. *Crossing the Threshold of Hope.* New York: Alfred
A. Knopf, 1995.

Kim, Ashida. *The Invisible Ninja: Ancient Secrets of Surprise.*
New York: Citadel Press, 1983.

Kim, Ashida. *Secrets of the Ninja.* New York: Citadel Press, 1981.

Lanning, Michael Lee and Dan Cragg. *Inside the VC and the NVA:
The Real Story of North Vietnam's Armed Forces.* New York:
Ivy Books, 1992.

Lind, William S. *Maneuver Warfare Handbook.* Boulder, CO:
Westview Press, 1985.

Lung, Dr. Haha: *Knights of Darkness: Secrets of the World's Deadliest
Night Fighters.* Boulder, CO: Paladin Press, 1998.

Maitland, Terrence and Peter McInerney. *Vietnam Experience:
A Contagion of War.* Newton, MA: Boston Publishing, 1968.

Maneuver Warfare: An Anthology. Edited by Richard D. Hooker, Jr.
Novato, CA: Presidio Press, 1993.

McPherson, James M. *Battle Cry of Freedom.* New York: Oxford
University Press, 1988.

The Military Art of People's War. Edited by Russel Stetler. New
York: Monthly Review Press, 1970.

Military History Online. As retrieved from its website, militaryhistoryonline.com.

Night Movements. Translated by C. Burnett. Port Townsend, WA: Loompanics Unlimited, n.d. Originally published as a Japanese training manual. Tokyo: Imperial Japanese Army, 1913.

Peterson, Kirkland C. *Mind of the Ninja: Exploring the Inner Power.* Chicago: Contemporary Books, 1986.

Phrases.org. As retrieved from the website of that name.

Poole, H. John. *Dragon Days: Time for "Unconventional" Tactics.* Emerald Isle, NC: Posterity Press, 2007.

Poole, H. John. *Gung Ho: The Corps' Most Progressive Tradition.* Emerald Isle, NC: Posterity Press, 2012.

Poole, H. John. *One More Bridge to Cross: Lowering the Cost of War.* Emerald Isle, NC: Posterity Press, 1999).

Poole, H. John. *Phantom Soldier: The Enemy's Answer to Firepower.* Emerald Isle, NC: Posterity Press, 2001.

Poole, H. John. *Tactics of the Crescent Moon: Militant Muslim Combat Methods.* Emerald Isle, NC: Posterity Press, 2004.

Poole, H. John. *Tequila Junction: 4th-Generation Counterinsurgency.* Emerald Isle, NC: Posterity Press, 2008.

Poole, H. John. *Terrorist Trail: Backtracking the Foreign Fighter.* Emerald Isle, NC: Posterity Press, 2006.

Poole, H. John. *The Tiger's Way: A U.S. Private's Best Chance for Survival.* Emerald Isle, NC: Posterity Press, 2003.

The Quotations Page. As retrieved from the following url: www.quotationspage.com.

Reid-Daly, Lt.Col. R.F. *Pamwe Chete: The Legend of the Selous Scouts.* Weltevreden Park, South Africa: Covos-Day Books, 1999.

Ross, Bill D. *Iwo Jima: Legacy of Valor.* New York: Vintage, 1986.

Soviet Combat Regulations of November 1942. Moscow: [Stalin], 1942. Republished as *Soviet Infantry Tactics in World War II: Red Army Infantry Tactics from Squad to Rifle Company from the Combat Regulations,* with translation, introduction, and notes by Charles C. Sharp. West Chester, OH: George Nafziger, 1998.

Stiff, Peter. *The Silent War: South African Recce Operations, 1969-1994.* Alberton, South Africa: Galago Publishing, 1999.

The Strategic Advantage: Sun Zi & Western Approaches to War. Edited by Cao Shan. Beijing: New World Press, 1997.

Strategy Page. As retrieved from its website, www.strategypage.com.

Sun Tzu. *The Art of War.* Translated and with introduction by Samuel B. Griffith. Foreword by B.H. Liddell Hart. New York: Oxford University Press, 1963.

TFP Student Action. As retrieved from its website, www.tfpstudentaction.org.

The Tibetan Book of Living and Dying. By Chogyal Rinpoche.
Torah (Old Testament of the Christian Bible).
The Travels of Marco Polo. Revised from Marsden's translation.
   Edited by Manuel Komroff. New York: Modern Library,
   1953.
Visuddhimagga (Path of Purification). By Buddhaghosa.
Warr, Nicholas. *Phase Line Green: The Battle for Hue, 1968.*
   Annapolis, MD: Naval Institute Press, 1997.
Warriors of Islam: Iran's Revolutionary Guard. By Kenneth Katzman.
   Boulder, CO: Westview Press, 1993.
Wikipedia Encyclopedia. As retrieved from its website, wikipedia.org.
Wright, Robin. *Sacred Rage: The Wrath of Militant Islam.* New
   York: Simon & Schuster, 1985.
Wukovits, John. *American Commando.* New York: New American
   Library (Penguin), 2009.
Yousaf, Brigadier Mohammad and Major Mark Adkin. *Bear Trap:
   Afghanistan's Untold Story.* South Yorkshire, UK: Leo Cooper, n.d.
Zabih, Sepehr. *The Iranian Military in Revolution and War.*
   London: Routledge, 1988.

Videotapes, Movies, DVDs, TV Programs, Slide Shows, CDs, Illustrations

"Brave Heart." VHS. 177 minutes. Paramount, 1995.
"Criminal Minds." Season 6, #602. *J.J. Episode.* ION TV, 9 December
   2013.
Documentary of house examination by supernatural investigators.
   Videotape. As viewed by author on educational TV. N.p.,
   n.d.
"An Entirely New War." Segment of *Korea—the Unknown War Series.*
   London: Thames TV in association with WGBH Boston, 1990.
   NC Public TV.
"Gung Ho: The Story of Carlson's Makin Island Raiders." Movie.
   87 minutes. Universal Studios, 1943. Based on factual story by
   Lieutenant W.S. LeFrancois USMC. With Lieutenant Colonel
   Evans Carlson USMCR as technical advisor.
"The Pacific." DVD. 10-part miniseries. 530 minutes. From Tom
   Hanks, Steven Spielberg, and Gary Goetzman. HBO and
   DreamWorks, n.d.
Price, Sergeant Toney. Two-hour interview by Kirk Hauser. Audio CD,
   2013.
"Sands of Iwo Jima." Videocassette. 109 minutes. Republic Pictures,
   1988.
"Sergeant York." DVD. 134 minutes. Warner Brothers Pictures.
   Isbn #1-4198-3829-6.

The Shadow. Audio CD. "The Making of a Legend." GAA Corporation, 1996. Stock#49030.

"We Were Soldiers." DVD. 84 Minutes. Warner Brothers, 2002. Based on the book, *We Were Soldiers Once and Young: Ia Drang—the Battle That Changed the War in Vietnam.* By Harold G. Moore and Joseph Galloway. New York: Presidio Press, 2004.

"Windtalkers." TV movie. 134 minutes. MGM, 2002.

"Wushu & Sanda." *Fight Quest.* Discovery Channel, 28 December 2008.

Letters, E-Mail, and Direct Verbal Conversations

Baldwin, Ron (member of CAC 10). In Amazon review of *Gung Ho,* 8 January 2013.

Bennett, Brad (81mm operator at CAC 10). In e-mail to author on 18 February 2013.

Curator of Iranian War Museum at Tehran. In conversation with author around 2002.

Dutch Marine exchange officer. In conversation with author on 16 January 2002.

Former Marine twin brother of Chief Warrant Officer in charge of indigenous Afghan military training. In conversation with author around 2011.

Guthrie, Tag (former squad leader in A/1/4). In multiple e-mails to the author from 2003 to 2014.

Hauser, Kirk (one of Sergeant Price's men). In e-mails to author in December 2013.

Highly experienced reconnaissance Marine who after retirement became training contractor for special-operations community. In an e-mail to author during January 2014.

Louie (Dan Marsh's son) at USMCRaiders.com. In e-mail to author of 9 December 2013.

Multi-tour U.S. Army Special Forces veteran of wars in Iraq and Afghanistan and expert mantracker. In e-mail conversations with author between June 2013 and May 2014.

Price, Sergeant Toney. Sketch of trail junction ambush, n.d.

Reinke, Doctor David H. (expert on parapsychology and Eastern religions). In telephone conversation with author in June 2001, and multiple e-mails thereafter.

Schmorde, Jim (Vietnam veteran of L/3/4). In e-mail to author of December 2012.

Seal Team Six member. In interview on CBS's "60 Minutes," n.d.

Newspaper, Magazine, Radio, and Website Articles

Collins, Joseph J. "Counterinsurgency & Common Sense." *Armed Forces Journal,* January/February 2013.

"Conventional Weapons: 30/11/2011 Overview." International Committee for the Red Cross. From the following url in March 2014: http://www.icrc.org/eng/war-and-law/weapons/conventional-weapons/overview-conventional-weapons.htm.

"Finland in the Winter War: An Examination of Finnish Infantry Tactics." By Ville Savin. N.p., n.d.

Gertz, Bill. "Notes from the Pentagon." *Washington Times,* 5 March 2004.

Griffin, Jennifer. "U.S. Intelligence Officer Reveals Secret Story of Saddam Hussein's Capture." FoxNews.com, 12 December 2008.

"Images for *Dac Cong.*" As retrieved from this url in March 2014: http://www.google.com/images?oe=UTF-8&gfns=1&q=dac+cong&hl=en&sa=X&oi=image_result_group&ei=r281U57HKbPQsQTqxIGgCg&ved=0CBsQsAQ.

Jelinek, Pauline and Anne Gearan (Associated Press). "General Seeks New Afghan Approach." *Philadelphia Inquirer,* 1 August 2009.

"Joseph Kony Hunt Is Proving Difficult for U.S. Troops." *Washington Post,* 29 April 2012. As retrieved in June 2012 from this url: washingtonpost.com/world/national-security/joseph-kony-hunt-is-proving-difficult-for-us-troops/2012/04/29/gIQAasM6pT_story_1.html

Kitchens, Lt.Col. Gregory H. USMCR. "Building the Team for Unit Excellence." *Marine Corps Gazette,* December 2003.

Krulak, Gen. Charles C. "The Strategic Corporal: Leadership in the Three Block War." *Marines Magazine,* January 1999. From this url: http://www.au.af.mil/au/awc/awcgate/usmc/strategic_corporal.htm.

Mao Tse-Tung. "Mao's Primer on Guerrilla War." Translated by B.Gen. Samuel B. Griffith. In FMFRP 19-9, *The Guerrilla and How to Fight Him.* Quantico, VA: Marine Corps Combat Development Command, 1990. And in *Marine Corps Gazette*, January 1962, plus a 1941 issue.

Morris. Loveday. "In Syria, Hezbollah Forces Appear Ready to Attack Rebels in City of Aleppo." *Washington Post,* 2 June 2013.

Moser, Don. "Their Mission Defend, Befriend." *Life Magazine,* 25 August 1967.

Munoz, Carlos. "Pentagon Taking New Tact in Terrorism Fight in Africa." *The Hill,* 14 July 2012.

Murphy, Dan. "Sadr the Agitator: Like Father, Like Son." *Christian Science Monitor,* 27 April 2004.

"Netherworld." As retrieved from the following url: http://chinesemythical.wordpress.com.

"Satan and Humanity." As retrieved from the following url in December 2012: http://www.islamawareness.net/Jinn/satan.html.

Scales, Maj.Gen. Robert H., U.S. Army (Ret.). "Infantry and National Priorities." *Armed Forces Journal,* December 2007.

"Sergeant Alvin York." By Dr. Michael Birdwell. Great War Society. As retrieved on 15 October 2009 from its website, www.worldwar1.com.

Simpson, Jim. "Scouts to the Rescue." *Defense Watch,* 17 September 2003.

"Stonewall Jackson." N.p., n.d. As retrieved from this url: sonofthesouth.net/leefoundation/jackson/battle-fredericksburg.html

Waterman, Shaun. "North Korean Jamming of GPS Shows System's Weakness." *Washington Times,* 23 August 2012.

"Why Did Armored Corps Fail in Lebanon." By Hanan Greenberg. Israeli News, 30 August 2006.

About the Author

After 28 years of commissioned and noncommissioned infantry service, John Poole retired from the United States Marine Corps in April 1993. While on active duty, he studied small-unit tactics for nine years: (1) six months at the Basic School in Quantico (1966); (2) seven months as a rifle platoon commander in Vietnam (1966-67); (3) three months as a rifle company commander at Camp Pendleton (1967); (4) five months as a regimental headquarters company (and camp) commander in Vietnam (1968); (5) eight months as a rifle company commander in Vietnam (1968-69); (6) five and a half years as an instructor with the Advanced Infantry Training Company (AITC) at Camp Lejeune (1986-92); and (7) one year as the Staff Noncommissioned Officer in Charge of the 3rd Marine Division Combat Squad Leaders Course (CSLC) on Okinawa (1992-93).

While at AITC, he developed, taught, and refined courses on maneuver warfare, land navigation, fire support coordination, call for fire, adjust fire, close air support, M203 grenade launcher, movement to contact, daylight attack, night attack, infiltration, defense, offensive Military Operations in Urban Terrain (MOUT), defensive MOUT, Nuclear/Biological/Chemical (NBC) defense, and leadership. While at CSLC, he further refined the same periods of instruction and developed others on patrolling.

He has completed all of the correspondence school requirements for the Marine Corps Command and Staff College, Naval War College (1,000-hour curriculum), and Marine Corps Warfighting Skills Program. He is a graduate of the Camp Lejeune Instructional Management Course, the 2nd Marine Division Skill Leaders in Advanced Marksmanship (SLAM) Course, and the East-Coast School of Infantry Platoon Sergeants' Course.

In the 21 years since retirement, John Poole has researched the small-unit tactics of other nations and written 13 other books: (1) *The Last Hundred Yards,* a squad combat study based on the consensus opinions of 1,200 NCOs and casualty statistics of AITC and CSLC field trials; (2) *One More Bridge to Cross*, a treatise on enemy proficiency at short range and how to match it; (3) *Phantom Soldier,* an in-depth look at the highly deceptive Asian style of war; (4) *The Tiger's Way,* the fighting styles of Eastern fire teams and soldiers; (5) *Tactics of the Crescent Moon,* insurgent procedures in Palestine, Chechnya, Afghanistan, and Iraq; (6) *Militant Tricks,* an honest appraisal of the so-far-undefeated *jihadist* method; (7) *Terrorist Trail,* tracing the *jihadists* in Iraq back to their home countries; (8) *Dragon Days,* an unconventional warfare technique manual; (9) *Tequila Junction,* how to fight narco-guerrillas; (10) *Homeland Siege,* confronting the 4GW assault by a foreign power's organized-crime proxies; (11) *Expeditionary Eagles,* how to outmaneuver the Taliban; *Global Warrior,* forestalling WWIII with tiny contingents; and *Gung Ho,* how supporting arms are not needed to take strongpoint matrices.

Since 2000, he has done research in Mainland China (twice), North Korea, Vietnam, Cambodia, Thailand, India (twice), Pakistan (twice), Iran, Lebanon, Turkey, Egypt, Sudan, Tanzania, Venezuela, and Sri Lanka. Over the course of his lifetime, he has visited scores of other nations on all five continents. When he tried to visit Lahore in the late Spring of 2011, his Pakistani visa request was not honored.

As of April 2014, John Poole had conducted multiday training sessions (on advanced squad tactics) at 41 (mostly Marine) battalions, nine Marine schools, and seven special-operations units from all four U.S. service branches.

Between early tours in the Marine Corps (from 1969 to 1971), he served as a criminal investigator with the Illinois Bureau of Investigation (IBI). After attending the State Police Academy for several months in Springfield, he was assigned to the IBI's Chicago office. There, he worked mostly on general criminal and drug cases.

Name Index

A

Ali ibn Abi Talib, 4th Caliph   51
Amos, Gen. James F.   35
Attila [the Hun]   9

B

Baldwin, Ron PFC   77
Barrett, Pvt. Carlton W.   195
Bennett, L.Cpl. Brad   78, 90, 91
Bill (Ryan's friend)   208
Bin Laden, Osama   244
Bryan, William Jennings   1
Burke, Edmund   14
Burke, S.Sgt.   64, 65
Bush, Cpl. Richard Earl   127,
    132, 134

C

Caesar   4
Carlson, B.Gen. Evans Forsythe
    13, 22, 36, 55, 93, 101, 102,
    112, 117, 120, 122, 123, 124,
    128, 132, 190, 201, 204, 237,
    239, 244
Chamberlain, Col. Joshua
    Lawrence   xxi, 84
Charteris, B.Gen. Charles   9
Christ (a.k.a. Jesus, Lord, or
    Redeemer)   3, 9, 242

Clausen, PFC Raymond Mike, Jr.
    102, 103, 104
Corbett, Col. Edward James   104
Corcoran, Capt. Richard   65
Cox, 1stLt. Jack "The Georgia
    Peach"   63
Creator (a.k.a. God)   xxvi, 3, 4, 7,
    9

D

Devil (a.k.a. Lucifer or Satan)   5,
    8, 12

E

Edson, Col. Merritt A. "Red Mike"
    198, 244

F

Flower, Ken   35

O

Obama, President Barack Hussein
45, 47

P

Petraeus, Gen. David H.   35, 36
Pinder, Tech. 5th Grade John J.,
   Jr.   196
Pol Pot, Gen. Secretary   31
Polo, Marco   15
Price, Sgt. Anthony Wayne "Toney"
   65, 77, 78, 79, 80, 83, 84, 90,
   91, 243
Puller, Lt.Gen. Lewis B. "Chesty"
   70, 84, 206, 243

Q

No entries

R

Redeemer (a.k.a. Lord, Jesus,
   or Christ)   3, 9, 242
Reinke, Dr. David H.   21
Ryan, Pvt. Robert B. "Squirt," Jr.
   208

S

Satan (a.k.a. Devil or Lucifer)   5,
   8, 12
Scales, Maj.Gen. Robert H.   67
Smuckatella, PFC Paul   209
Stein, Cpl. Tony   83, 235, 236,
   237
Stiff, Peter   35
Sun Tzu   41, 48, 112

Sutton-Price, Ted   34
Sykes, Maj.   15

T

Tamerlane   36
Thomason, Sgt. Clyde A.   116
Trish (Ryan's prospective wife)
   208

U

No entries

V

Verrier, Anthony   34

W

Walsh, Gy.Sgt. William Gary
   127, 128, 131, 132
Walt, Gen. Lewis William   36
Westmoreland, Gen. William C.
   27

X

No entries

Y

York, Sgt. Alvin C.   3, 4
Yousaf, Brigadier Mohammad
    53, 54
Yule, Col. Henry   15

Z

Zvobgo, Edison   34